Participatory Media in Environmental Communication

AF849538

Participatory Media in Environmental Communication brings together stories of communities in the Pacific Islands – a region that is severely affected by the impacts of climate change. Despite living on the margins of the digital revolution, these island communities have used media and communication to create awareness of and find solutions to environmental challenges. By telling their stories in their own way, ordinary people are able to communicate compelling accounts of how different, but interrelated, environmental, political and economic issues converge and impact at a local level.

This book fills a significant gap in our understanding of how participatory media is used as a dialogic tool to raise awareness and facilitate discussion of environmental issues that are now critical. It includes a section on pedagogy and practice – the undergirding principles, the tools, the methods. The book offers a framework for Participatory Environmental Communication that weaves three widely used concepts, *diversity, network* and *agency*, into a cohesive underlying system to bring scholars, practitioners and diverse communities together in a dialogue about pressing environmental issues.

This book is a valuable resource for researchers and students in communication and media studies, environmental communication, cultural studies, and environmental sciences, as well as practitioners, policy makers and environmental activists.

Usha Sundar Harris is an academic in the Department of Media, Music, Communication and Cultural Studies at Macquarie University, Australia.

Routledge Studies in Environmental Communication and Media

Environmental Communication and Critical Coastal Policy
Communities, Culture and Nature
Kerrie Foxwell-Norton

Climate Change and Post-Political Communication
Media, Emotion and Environmental Advocacy
Philip Hammond

The Discourses of Environmental Collapse
Imagining the End
Edited by Alison E. Vogelaar, Brack W. Hale and Alexandra Peat

Environmental Management of the Media
Policy, Industry, Practice
Pietari Kääpä

Participatory Networks and the Environment
The BGreen Project in the US and Bangladesh
Fadia Hasan

Participatory Media in Environmental Communication
Engaging Communities in the Periphery
Usha Harris

For more information about this series, please visit: www.routledge.com/Routledge-Studies-in-Environmental-Communication-and-Media/book-series/RSECM

Participatory Media in Environmental Communication

Engaging Communities in the Periphery

Usha Sundar Harris

Routledge
Taylor & Francis Group
LONDON AND NEW YORK

earthscan
from Routledge

First published 2019 by Routledge

2 Park Square, Milton Park, Abingdon, Oxon, OX14 4RN
605 Third Avenue, New York, NY 10017

Routledge is an imprint of the Taylor & Francis Group, an informa business

First issued in paperback 2020

British Library Cataloguing-in-Publication Data
A catalogue record for this book is available from the British Library

Library of Congress Cataloging-in-Publication Data
A catalog record has been requested for this book

ISBN: 978-1-138-65528-7 (hbk)
ISBN: 978-0-367-73356-8 (pbk)

Typeset in Sabon
by Apex CoVantage, LLC

To the people of the Pacific Islands for their generosity and resourcefulness in the face of mounting natural disasters.

Contents

Illustrations

Figures

Tables

Case studies

List of boxes

Acknowledgements

In a project such as this, many people are deserving of my great appreciation. I owe debts of gratitude to many in the Pacific, all of whom have demonstrated grace and generosity in opening their homes and communities to me and sharing their wisdom and knowledge that has informed the contents of this book. They have taught me the values of reciprocity, goodwill and community networking. Particular appreciation is extended to Rex Thomas in Vanuatu, Rommel Sharma in Fiji, Claire Antarea in Kiribati and the dedicated staff at Pacific Calling Partnership for introducing me to community organisations and decision-makers.

I would like to acknowledge the support of *Pacific Journalism Review* and editor David Robie for publishing an earlier version of chapter three titled "Engaging Communities in Environmental Communication". Similarly, chapter six is a revised version of an article published in the special issue of *Information Technologies & International Development* co-authored with Evangelia Papoutsaki and Sandra Kailahi.

The book has greatly benefitted from advice and feedback provided by Evangelia Papoutsaki. She has been a true friend and mentor over the years. My immense appreciation to Mehdi Azam, my research assistant, who dedicated many hours creating and cross-checking references, conducting literature search and providing assistance with layout. Thank you to various others who have assisted in the draft and production process: my colleague Nicole Matthews for her comments on the chapter two draft; Annabelle Harris and Matt Shobbrook at Routledge, for their persistence and patience in guiding the project; Macquarie University Faculty of Arts for the mid-career grant which funded teaching buyout and a research assistant.

Gratitude is due to my family for their absolute support and belief in me. Reg Harris, my husband and greatest supporter, for the long hours of proofreading and valuable comments on the drafts; my son Reg Jr for his audio expertise in recording the interviews in Vanuatu and transcribing; and daughter Anjali for her words of inspiration. Special thanks are also due to my aunt, Kamla Jokhan, who has generously looked after me during my many visits to Fiji.

This book is dedicated to my late parents, Ram Sundar and Pranpati, who encouraged me to find my passion and discover my potential. They gave me my first lessons in community empowerment.

Abbreviations

A2C2	Action Against Climate Change
ANT	Actor Network Theory
BRA	Bougainville Revolutionary Army
C4D	Communication for Development
CAN	Climate Action Network
CDSC	Communication for Development and Social Change
CIVICUS	World Alliance for Citizen Participation
COP	Conference of the Parties
CSO	Civil Society Organisations
DAPL	Dakota Access Pipeline
DDT	Dichlorodiphenyltrichloroethane
DNA	Diversity, Network and Agency
DRR	Disaster Risk Reduction
EEZ	Exclusive Economic Zone
ENGO	Environmental Non-governmental Organisations
ENSO	El Niño Southern Oscillation
FAO	Food and Agriculture Organisation
GCF	Green Climate Fund
GHGs	Greenhouse Gases
GIZ	German Agency for International Cooperation
ICT	Information and Communication Technology
IECA	International Environmental Communication Association
IPCC	Intergovernmental Panel on Climate Change
ITU	International Telecommunication Union
IUCN	International Union for Conservation of Nature
KiriCAN	Kiribati Climate Action Network
LAN	Local Area Networks
LGBTQI	Lesbian, Gay, Bisexual, Transgender, Queer or Questioning, and Intersex
MIT	Massachusetts Institute of Technology
NFIP	Nuclear Free and Independent Pacific

NGO	Non-governmental Organization
NSW	New South Wales
OCHA	United Nations Office for the Coordination of Humanitarian Affairs
PACMAS	Pacific Media Assistance Scheme
PAR	Participatory Action Research
PCC	Pacific Conference of Churches
PCP	Pacific Calling Partnership
PEC	Participatory Environmental Communication
PIC	Pacific Island Countries
PICAN	Pacific Island Climate Action Network
PLA	Participatory Learning Action
PM	Participatory Media
PNG	Papua New Guinea
PRA	Participatory Rural Appraisal
PRIF	Pacific Region Infrastructure Facility
PV	Participatory Video
RTZ	Rio Tinto Zinc
SD	Secure Digital
SGI	Soka Gakkai International
SMS	Short Message Service
SPC	Secretariat of the Pacific Community
SPREP	Secretariat of the Pacific Regional Environment Programme
STS	Science and Technologies Studies
TC	Tropical Cyclone
TCAN	Tuvalu Climate Action Network
TEK	Traditional Ecological Knowledge
TK	Traditional Knowledge
UGC	User-generated Content
UN	United Nations
UNDP	United Nations Development Programme
UNEP	United Nations Environmental Programme
UNESCO	United Nations Educational, Scientific and Cultural Organisation
UNISDR	United Nations International Strategy for Disaster Reduction
UN-OHRLLS	United Nations Office of the High Representative for the Least Developed Countries, Landlocked Developing Countries and Small Island Developing States
VCAN	Vanuatu Climate Action Network
WCED	World Commission on Environment and Development

Part I

Theory in practice

1 Who is telling our stories?

Reflection

It was the evening of 19 February 2016. A category 5 Tropical Cyclone named Winston headed directly for the Fiji Islands. Winds of up to 325 kilometres per hour left a trail of destruction as Winston battered coastal villages, rural towns and luxury resorts. On the Internet, photographs, videos and stories filtered in through different social media platforms, including Twitter, Facebook and Instagram. People told their stories in 16 characters or less, and through their microblogs a narrative emerged about humanity's utter vulnerability against the force of Nature. Terrified families tried to find safety inside cupboards, under beds and kitchen benches as roofs and walls were blown away by the strongest cyclone recorded in the southern hemisphere. The most compelling story to emerge that night was in the form of a twitter feed from a young woman caught in the eye of the storm in a small rural town in the north of Viti Levu, Fiji's main island. She took safety in a cupboard with her family as the walls of her house collapsed around them. Her tweets, disarming in their honesty, provided vignettes of her family's harrowing encounter with TC Winston as it destroyed their family home. Those tweets appeared in newspapers all around the world (NZ Herald, 2016):

> The house is leaking and everyone's crying. Planning where to hide when the entire roof comes off.
>
> This is hands down the scariest point of my life.
>
> Everyone's safe in the cupboard with a mattress against us, no roof, barely any house, but alive so far. Thank God.

The next day TC Winston tore across the remote Koro Island, where families experienced sheer terror as cyclonic winds whipped up tsunami-like storm surges destroying 14 villages. When it finally left the Fiji Islands, the category 5 cyclone had killed 40 people and impacted 40% of Fiji's population, who were either homeless or scarred by its fury (OCHA, 2016).

The stories about Cyclone Winston helped to revive my own childhood memories of natural disasters in Fiji. Tropical rains, humidity, hurricanes and floods are regular occurrences of Pacific island life and people name the seasons accordingly – things happen either in cyclone season or out of cyclone season. Growing up in the river town of Nausori with our house situated on the high banks of the Rewa River, I witnessed numerous cyclones, but their strength never exceeded category three; the strongest of these, in my memory, being Hurricane Bebe in 1972. After the hurricanes came the inevitable floods. My mother, who worked as a Red Cross volunteer, would enlist the support of local businesses, family and friends for donations of food, clothes and cash during these times of disaster. She would save sheets, blankets and clothes in big woven baskets for such events and coax us to donate our almost new dresses with reassuring words that these would bring solace to children who had lost everything. With other Red Cross volunteers, she would then spend days distributing the supplies to people in the most vulnerable areas. Upon her return, we would hear stories of the sheer hopelessness of mothers who could not feed or clothe their children, and of despair on the faces of men whose only means of income were the cash crops now destroyed in floods. There were no cameras or smartphones to capture their plight, but the stories moved our hearts.

Introduction

Humans share their joy, sorrow, knowledge and experiences with others through storytelling. Stories are the way in which we communicate our past and our present to those around us and to our future generations. Our ability to pass on collective knowledge from one generation to another has been an important element of human progress and dominance over the natural environment. Unlike any other species, humans have the capacity to record and accumulate knowledge in the form of images and words to pass on to the next generation. This ability of our species to create, record and improve upon a collective pool of knowledge has progressed us from cave drawings to the digital age that we experience today.

David Christian, in his book *Maps of Time: An Introduction to Big History* (2011: 145), observes "the transition to human history is primarily marked not by a change in the nature of humans as individuals but rather by a change in the way individuals relate to each other". This ability to 'learn collectively' (Christian, 2011: 146) and share knowledge using our capacity for symbolic language has made humans both smart and adaptable. By accumulating a growing pool of knowledge passed down from one generation to the next, through cooperation and collective action, we have improved our survival skills, giving us the ability to live in one place, cultivate food crops and develop thriving civilisations.

It is people's capacity to tell stories that provides a society with the means to remember the past, understand the present and plan for the future. "Culture lives because it is communicated" (Gumucio-Dagron, 2014: 110). Technologies may have changed over time but one thing has remained constant – our need to share stories. Communication in its many forms, from early cave drawings and oral traditions to sophisticated forms of language, written texts and digital technology, has contributed significantly to the repository of knowledge and its dissemination through processes of socialisation. Spurgeon describes storytelling as a "timeless human tool for both the reinforcement of ideas and values and the emergence of new ones" (2015: 133). Nisbet and Scheufele (2009: 1775) identify the storytelling technique as "an important mechanism" in attracting diverse audiences to improve public engagement in science communication.

Storytellers

Storytellers have intimate knowledge of their audience and the spaces they inhabit. Myths and legends include familiar places, cultural artefacts and characters inspired by the surrounding flora and fauna. Context informs content, which embeds knowledge and skills crucial to a group's survival. The dot painting and oral traditions of the Australian Aborigines embed knowledge about water and food sources, campsites and ceremonial places (Berndt and Berndt, 1964; Lawlor, 1991). Lawlor explains the Aboriginal women's sand drawing: "Many symbols drawn with a story are created on the spot, to fit a particular story and a particular audience. The storyteller fabricates and defines her symbols as she communicates them, which makes all symbols highly contextual" (1991: 285). Contemporary artist Ben Fong uses similar place-based environmental education to teach children on the island of Bega in Fiji about the dangers of plastic to marine life. By drawing a fish in the sand and placing plastic bags collected from the beach in the belly of the fish, he creates a visual reference point about environmental degradation and the importance of proper rubbish disposal.

Traditional societies communicate their ecological knowledge by means of cultural practices. The seed keepers of the Matigsalog tribe in Marilog, Davao City, Philippines provide an example of how women have safeguarded resilient varieties of rice and passed these down to the next generation. The women bequeath the 'heirloom rice' as dowries to their daughters and at the same time impart knowledge of seed bank, farming practices, and their healing powers (Cabusao, 2012). In Cordillera region of the Philippines 'heirloom rice' has been preserved and harvested by women to ensure their family's future food security:

> After being handed down in an unbroken link, from generation to generation, more than 300 of these native rice varieties achieved a venerated status as tribal heirlooms. Heirloom rice is a spiritual bridge to the ancestors who built considerable knowledge through trial and error and fashioned unique technologies from experiences collected over the centuries. It has become as much a part of the region's culture and identity as the resplendent rice terraces that the people's forebears carved out of the mountainsides.
>
> (Santiaguel, 2010)

This enables the next generation to carry the historical and cultural memory of food production and preservation techniques and to adapt their practice in future as circumstances change. Similar practices are found in other parts of the world where women are the knowledge bearers of food preservation, biodiversity and cultivation practices (Lopez, 2012).

Storytelling – the means by which we share knowledge and emotion – is at the heart of this book and the style it adopts. The book integrates a personal point of view based on my life in Fiji where I grew up and worked as a journalist, and later, as this experience translated into my research interest in media and communication in the Pacific. This convergence of experiential knowledge gathered as a media practitioner and practice-based researcher with communities in the Pacific context provide the insights which I bring into the present discussion. Working in both mainstream and community media settings, I have come to understand how proactive use of media can disrupt power relations and become a social and political force for marginalised groups. By working outside the established norms and networks, community producers create alternative dialogic spaces for self-expression. The narrative form connects people in a meaningful way by highlighting their struggles and adversities. This capacity for self-representation leads them to reflect on the power relationships "within their specific social structures and cultural contexts", as well as those from outside (Braden, 1999: 118).

Tradition and Fiji COP23

Storytelling is important to both the maintenance of culture and everyday interaction in Pacific island communities. As such it is an integral part of the island communication ecology. During Fiji's Presidency of the annual United Nations climate change conference known as Conference of Parties (COP) 23 in Bonn, Germany in 2017, Fiji's Prime Minister, Frank Bainimarama, arrived bearing three symbolic gifts from the Pacific. The gifts were the *Bula* spirit, the Fijian greeting which symbolises inclusiveness, friendliness and solidarity, *drua*, a traditional canoe, and *talanoa*, an indigenous form of dialogue practised in the Pacific. An important Pacific icon, the *drua* was placed at the entrance of the main hall to serve as a powerful symbol of unity to remind conference attendees that there is only one earth and we are all in the same boat. *Talanoa* became the guiding method of negotiation which Bainimarama explained in the following way:

> This is a process of inclusive, participatory and transparent dialogue that builds empathy and leads to decision making for the collective good. It is not about finger pointing and laying blame but is about listening to each other, learning from each other, sharing stories, skills and experiences.

The United Nations has adopted the *talanoa* process in climate change discussions leading to COP24 in Poland. "Countries will follow the traditional Fijian practice of *talanoa* in their negotiations – a way of solving problems by sharing ideas, skills, and experience through storytelling", *The Times* in UK reported under the headline, "Fijian storytelling can save world, says UN" (Webster, 2017).

This is an example of how alternative discursive spaces and indigenous ways of knowing can gain greater acceptance within conventional knowledge systems and communication practices. The International Union for Conservation of Nature (IUCN) report recommends the inclusion of affected communities "in the international, regional and local climate change discourse" (Macchi, 2008: 59). It is increasingly clear now that diverse forms of knowledge – inclusive of scientific, experiential and traditional – are essential in order to understand vulnerabilities and build resilience of local communities in the face of environmental changes and to identify effective adaptation methods on the ground.

Great challenges posed by anthropogenic climate change and environmental degradation require human beings to rethink our own place on this planet and find a different way of living. A social and cultural transformation is necessary before we find an economic model that is not dependent on infinite growth, a model that does not promote a culture of insatiable consumption that depletes and degrades planet Earth's resources. Our relationship with Earth and our place within it requires a transformational shift in our own thinking. What role can environmental communication play in bridging the culture and nature divide in this current epoch described as the Anthropocene, during which human activity and the global economic system has had significant impact on Earth's system?

Gaps in environmental communication

Environmental communication is an emerging discipline within communication studies. The field of study has grown rapidly as governments, scientists, media, the private sector and civil society groups recognise the crucial role of effective communication about critical environmental concerns that affect both the natural environment and human society. The leading research and professional network, International Environmental Communication Association (IECA), describes the area as "a lay activity and a field of professional practice" and gives a functional definition of the field on its website:

> In the simplest terms, environmental communication is communication about environmental affairs. This includes all of the diverse forms of interpersonal, group, public, organizational, and mediated communication that make up the social debate about environmental issues and problems, and our relationship to the rest of nature.
>
> (Meisner, 2015)

One of the leading textbooks in the area, *Environmental Communication and the Public Sphere* (Cox and Pezzullo, 2016: 13) defines environmental communication as a form of symbolic action: "as the pragmatic and constitutive modes of expression – the naming, shaping, orienting and negotiating – of our ecological relationships in the world, including those with nonhuman systems, elements and species". In another articulation of environmental communication, Lie and Servaes (2015: 251) see it as a relatively new sub-discipline within communication science that overlaps with environmental education and health communication, especially in its focus on public engagement and public opinion. The interdisciplinary nature of the field means that environmental communication is also an important area of discussion in environmental science, environmental planning, development studies, and disaster risk management, among other areas. Climate change communication is a sub-theme that continues to dominate environmental communication.

In 2004 Alexander Flor at the University of Philippines Open University published a slim textbook titled *Environmental Communication: Principles, Approaches and Strategies of Communication Applied to Environmental Management*. It was based on Flor's own environmental practice and teaching experience of 25 years and was intended for use by graduate students of communication and environmental science. Flor (2004) emphasised the interrelation and interconnectedness of ecology as the undergirding principle in environmental communication. He was inspired by *The Closing Circle* (1971), written by American biologist Barry Commoner. Flor's book provides insights into the nature culture synergy:

> Environmental communication is founded in our culture, specifically, in traditional cultures. We see this in the environmental wisdom handed down from one generation to another; in the meanings that we attach to seasons; and in the reverence that our traditional cultures attach to farming and other forms of natural resource exploitation. Nature and culture are seen not as dichotomies but as a continuum. Hence, environmental communication programs should always contain a cultural dimension.
>
> (2004: 18)

In this definition, environmental communication is a way of talking about, learning from and disseminating information and knowledge about the natural environment from diverse knowledge holders inclusive of indigenous knowledge and non-human knowledge of other species embedded in their behaviour patterns.

Environmental communication research and analysis have centred on the top-down dissemination model which uses mainstream media or strategic communication to disseminate messages from policy makers, scientists and other interest groups to the wider population. Another research area analyses media reporting with emphasis on media framing and the attention that media gives to certain issues to the exclusion of others, especially in the sub-field of climate change communication (see Farbotko, 2005; Boykoff and Boykoff, 2007; Hulme, 2009; Carvalho, 2010; Schäfer et al., 2012; Zhao, 2009). Less frequent are studies that expand our understanding of the different communication channels' roles, as Olausson observes, "in shaping how citizens understand climate change" (Olausson, 2011: 282; Falaki and Adegbija, 2013). In other words, research has largely looked at media content and not audience reception of the content or engagement in content creation.

The task of tackling impacts and risks of climate change needs to be considered in new, creative ways that involve effective collaboration and whole-of-community participation. Anderson (2015: 380) urges scholars to move beyond the deficit model: "the idea that all we need to do is educate the public better". She argues that environmental communication must move away from a mediacentric approach which produces a "partial and narrow picture" and fails to "shed light on the wider cultural politics of environmental issues". In a similar vein, Nisbet and Scheufele (2009: 1768) posit that the 'deficit' model assumes science literacy to be the magic bullet whereby "facts are assumed to speak for themselves and be interpreted by all citizens in similar ways". Instead, they suggest science communication should facilitate dialogue with various stakeholders and "recognize, respect, and incorporate differences in knowledge, values, perspectives, and goals".

Meanwhile, to find new directions in research focus, Anders Hansen (2015) presents a strong case for an integrated approach to studying the three core areas of communication research – production, content and audiences. He challenges environmental communication researchers to "reconnect, empirically, and within clearly articulated theories and models of communication processes, the study of production, the content, and the social/political implications of environmental communication" (Hansen, 2015: 389). In Hansen's view these three areas highlight how "communicative 'power' in society is deeply unequally distributed" (2015: 389). This is because people's ability to "participate in and influence the nature of public 'mediated' communication about the environment" is dependent upon economic, political and cultural power (Hansen, 2015: 390). Participatory media is one way of bringing the voices in the margins into the environmental debate, thereby disrupting the centres of media production.

Mass communication technologies have displaced traditional modes of information sharing such as storytelling, song, dance and folk theatre in many societies. In the 20th and 21st centuries, the stories that inhabit our imagination are of cultures that own the communication infrastructure and have access to media channels. The printing press, radio and television have centralised the production and dissemination of information in the metropolitan centres, thereby concentrating knowledge and subsequently power into the hands of the few. National and international issues and events have to a great degree replaced local concerns in the news media.

Entertainment is now the preserve of a handful of multinational corporations. Curtin, Holt, and Sanson (2014: 5) in *Distribution Revolution: Conversations about the Digital Future of Film and Television* contend that "triumphant technologies throughout history have either been fostered by powerful interests or are ultimately put to work on behalf of influential elites". Television remains the most impenetrable of these media systems with its pursuit for high-end production values and mass audience appeal. The producers of one-way mass media work behind inscrutable codes of practice, professional standards and programme quality, which have placed conventional media beyond the reach of ordinary people. While limited spaces for audience participation are available through letters to the editor columns and talk-back radio programmes, these are allowed within the boundaries stipulated by producers who set the terms for such participation. The ability to participate is limited by one's access to technology, literacy skill and the prevailing political and cultural environment. In the struggle to capture public attention, political and economic elites dominate the mainstream media agenda that largely excludes voices at the margins.

At this point it is worth considering a question posed by Cox (2007: 6–7): "What is the *purpose* of environmental communication as a field?" Here I find Flor's (2004: 4) definition of environmental communication useful: "it is the deliberate exchange of environmental information, knowledge, and even wisdom". The exchange of knowledge rather than transmission of information is what engages people. Through dialogue and discussion people build their capacity to identify the links between environmental changes and impacts on their livelihoods, access to energy, food and water security, health and sanitation, among other issues.

In this regard, environmental communication can benefit from experiences in the communication for development field which has more than three decades of literature on alternative communication paradigms. Scholars and practitioners in the field of communication for development (see Melkote and Steeves, 2001; Hemer and Tufte, 2016b) engaged in a similar debate in the 1980s when it became clear that the dominant paradigm had major deficiencies. In that paradigm, mass media was seen as the magic bullet that would spread messages about development. The past mistakes of the early development models hold important lessons for environmental scholars and practitioners who want to find approaches other than the one-way, top-down dissemination of messages, also referred to as the deficit model (Brulle, 2010; Anderson, 2015).

The evolution of participatory communication offers important insights to those working in the growing environmental communication field. Decentralised horizontal communication models which emerged in the 1980s focused on the receiver instead of the sender, and valued the cultural identity of communities. Instead of the top-down monologic mode of information transmission, the new participatory approach gave rise to dialogic or two-way communication processes between decision-makers and citizens, giving voice and choice to the people who were at the receiving end of the development agenda – farmers, women and indigenous people (Servaes, 1999). It placed people and their culture at the centre of development, as Hemer and Tufte observe:

> With the focus shifting from the development state to the local community, communication became more than simply an instrument for persuasion and individual behaviour change; it was increasingly regarded as a process of democratization and empowerment and hence an end in itself.
>
> (2016a: 15)

In this framework scholars see participation as a basic human right and argue for a model where participation is seen as an end in itself rather than as a means to an end (Díaz Bordenave, 1989; White et al., 1994; Melkote and Steeves, 2015). People must acquire the knowledge and ability to develop their own communities.

Participatory environmental communication – an emerging framework

With an obvious need to find new ways of thinking about and practising environmental communication, I offer a conceptual framework for

Participatory Environmental Communication (PEC). In introducing the term Participatory Environmental Communication, I bring together epistemic knowledge and practice-based insights from the two areas of participatory communication and environmental communication.

PEC incorporates three interrelated strands: *Diversity, Network and Agency*. I refer to these as the *DNA* in participatory environmental communication because each of these strands is important to the survival of both natural and social systems. They are the essential building blocks of a resilient society. Embedded in these elements are the core principles of participatory communication which White and Patel so eloquently describe as "equalitarian, transactive and dialogic" in nature (White et al., 1994: 363). Participatory environmental communication is a process-oriented approach, which engages ordinary people in dialogue about environmental concerns so that they are able to identify the problems and are collectively empowered to make decisions to improve their situation. The term 'ordinary people' in this book is used in contrast to elite voices and powerful interest groups that frame the discourse in mainstream media.

By offering an alternative framework for analysis and practice of environmental communication, I am repositioning and rearticulating ideas and concepts that scholars have previously contributed to the broader field of communication. As such, the framework is offered not to the exclusion of other equally viable frameworks, but simply to add an alternative way of engaging in environmental communication by nurturing collaboration, dialogue and participation of communities in the periphery. The *DNA* framework brings together my experiential knowledge of participatory media and community engagement in the context of the Pacific region, one of the first places to feel the impacts of climate change. An extended discussion of the framework is provided in chapter 2.

What is participatory media?

The term 'participatory media' is used to emphasise the pivotal role of participation of lay people in making media content which is of special value to them and their community. Its value resides in the inherent purpose for which the content was created. The process of production invites collaboration, dialogue and mutual problem-solving, whereby people gain critical understanding of the problem and what action they need to take to solve it (Kheerajit and Flor, 2013). The communication between participants empowers endogenous networks and uncovers unique solutions to vexing problems arising from a particular context – be it historical, social, cultural, economic, political or geographic – of the community. In doing so it embeds local knowledge and finds solutions from within.

The reason I have chosen to use the term participatory media instead of alternative media, citizen's media or community media (see Atton, 2015 for explanation of each) is that the term crosses interdisciplinary boundaries more easily as environmental humanities, environmental science and climate change scientists adopt participatory approaches in their own work. Participatory media provides an important platform for communities to share knowledge and create awareness about environmental issues providing their own perspective. During times of crisis it becomes a critical source of information when mainstream media cannot get to a site or has no interest in the crisis. In this interpretation my use of participatory media aligns closely with the definition of 'citizen's media' given by Clemencia Rodríguez:

> In sum, alternative and community media research should re-centre the agency of communicators over technologies and refocus on context, uses and needs . . . as communities experience shifting contexts because of climate change, health crises or political uprising and protests, communication and media ecologies, needs and uses shift and change.
>
> (2016: 36–37)

Fenton concurs and proposes "not beginning with the media at all, but rather starting with context" (2016: 10).

Participatory media is defined as media culture that enables ordinary people to collectively identify problems, gather information, analyse, design and share media content which has value to them and their network. Participatory media is not solely about accessing technology. It is about ways in which participatory platforms enable people to have a voice, to organise and to demand change. Ordinary people who are excluded from participating in top-down hierarchical communication systems, be it village meetings or mainstream media, where they are only passive receivers of information, become active agents of change as they gain power over their own storytelling. The process of dialogue which engages diverse networks in finding new solutions to old problems catalyses transformative thinking by their active engagement in media production. Here, the boundary between the producer and consumer of content disappears.

A mix of mainstream media, strategic communication and access to participatory media technologies is vital in environmental communication. Mainstream media is critical during emergencies and the onset of natural disasters, strategic communication is important in education and behaviour change while participatory media engages communities in local problem solving. Each has its own place and purpose in society. In my own research in Fiji, I found that rural women who had engaged in participatory video production viewed foreign television programmes with a critical eye. A rural woman in Fiji can be engaged in participatory video production while at the same time enjoying the machinations of the protagonists in an imported soap opera produced in the United States for an American audience. That woman now has become a savvy and critical reader of the same information because of her increased media literacy (Harris, 2008).

Typology of participatory media

For the purpose of this book a typology of participatory media is provided below – classified into three sub groups – traditional and folk media, conventional media, and digital media. Each of these may function contemporaneously and the choice is influenced by the nature of the message, the participants and the society in which it is implemented. This approach has a close bearing to Downing's (2001, 2011) definition of alternative media which is analogous with a tapestry. To clarify their uses in this book each subgroup is defined, including the forms of media found in each.

Traditional and folk media

Traditional and folk media forms include face-to-face communication in either small or large groups such as oral narratives, performance and arts and crafts. Examples include the use of storytelling in the form of theatre, dance, puppetry, music and visual arts and craft. Oral narratives embed the community's intimate knowledge of its environment; how to navigate the oceans by the stars, or how the water flows through certain channels and how to harvest it. These stories then become a community's traditional ecological knowledge.

Conventional media

These include media forms which were popular before the digital era. Radio, television, newspapers, magazines, books, billboards and pamphlets are examples of conventional analogue media. Participatory media in this form included the use of community-based analogue video and radio production. In the 1970s, as more accessible video formats entered the market, activists saw these as powerful tools to disseminate messages of reform. Communities around the world realised the power of video and incorporated it into their everyday struggles for cultural preservation, representation, dialogue and activism with the invention of portapaks (self-contained portable video systems).

Before these interventions, filmmakers had told powerful stories about the intricate relationship between humans and the natural environment. These early film-making experiences captured the growing consciousness of voices in the margins. John Flaherty's *Nanook of the North* (1922) was not only the first documentary, but the first film that captured the wilderness and human's place in it, despite the controversy about Flaherty's staging of scenes. The film tells the story of the struggle of an Inuit hunter, Nanook, and his family to survive in the harsh Arctic environment of northern Quebec, Canada. In 1926 Flaherty released *Moana*, a documentary he filmed in Samoa. The story, centring on initiation to manhood, was filmed in Safune on the island of Savai'i where he spent more than a year with his family.

An influential series of films was produced by early pioneers, Colin Low and Donald Snowdon, for The National Film Board of Canada's '*Challenge for Change*' project. They laid down the epistemological foundation of participatory video in the *Fogo Process* in which the process of production was used as an empowerment tool for community development in the Fogo Island project in Newfoundland. The films deal with themes of displacement and connections with the land, not unlike those faced today by groups displaced by dam construction in India and China. Since then, process video in its many incarnations has been used around the world and has emerged as a sub-field in communication for development study and practice (see Milne et al., 2012; White, 2003).

Video has empowered indigenous communities in Australia and the Amazonian rainforest, and has provided a voice of resistance for Favela dwellers in Brazil and rural women in many parts of the world (see Gumucio-Dagron, 2001). The Food and Agricultural Organisation's (FAO) use of video[1] for agricultural extension work is well recognised (Lie and Mandler, 2009). From analogue formats, video and audio have become digitised and hybridised as a result of technology convergence. Amateurs who once needed a video camera, a recording deck, sound equipment and lights, can now shoot high quality video using their pocket smartphones and post these to social media sites instantaneously.

Digital media

Digital media, also referred to as new media, evolved after the convergence of telecommunications, broadcasting and Internet technologies, which eventually gave rise to web 2.0 and interactivity. Easy access and use of digital media has resulted in a paradigm shift from producer-oriented, top-down, capital-intensive content creation to user-generated content using peer-to-peer networks.

A new form of participatory culture has emerged. This includes mobile technologies such as smartphones and Internet-based social media. The evolution of web 2.0 and the resulting social media platforms provide for more participatory and interactive engagement – lowered production cost and higher participation in content sharing (Rheingold, 2008a, 2008b). Web 2.0 services focus more on socially connected web applications such as wikis, blogs, photo-video sharing, community media, participatory video, social networking and podcasts where the online presence of the users is important (Andersen, 2007). User-friendly web 2.0 applications thus allow active participation of end users, both as consumers and as prosumers, who require very limited technical knowledge compared to conventional media (Harrison and Barthel, 2009). People can also form online communities and communicate within networks through Facebook, Twitter, YouTube, Google+, Instagram and other emerging social media platforms.

According to Jenkins (2006: 2–3), media convergence means "the flow of content across multiple media platforms, the cooperation between multiple media industries, and the migratory behavior of media audiences who will go almost anywhere in search of the kinds of entertainment experiences they want". Media convergence relates to participatory culture and collective intelligence, thus representing a cultural shift rather than just a technological shift where consumers willingly participate in searching out new data, and making interactions among scattered media content (Jenkins, 2006). Convergence involves a change in the way media is produced and consumed. Such participatory culture, using web 2.0 applications and communication technological development, has transitioned the user from passive consumer to active producer of content in the social media platform.

Mobile technology

Smartphone technology has accelerated participatory culture by allowing the individual to create content using web 2.0 applications with a simple smartphone. The invention of the smartphone makes it possible for an individual to connect with the Internet anytime from anywhere. Many rural and remote communities are now fully connected with the help of the 3G/4G/LTE enabled smartphone (Chi and Lai, 2015). Improved access in lower income countries renders smartphones a necessity in the participatory culture of the Global South.

Easy access gives users the ability to connect, update, contribute and publish content using a variety of applications (apps). This also allows grassroots participation and activism on environmental issues. Environmental activist organisations such as 350.org and GetUP largely rely on online activism in their campaigns. Self-publishing on YouTube using smartphones equipped with video cameras, and micro-blogging sites such as Twitter, allow on-the-spot coverage of disasters by people who share their stories and pictures, thus transforming the public sphere (Cameron, 2007). At this time, digital participatory culture remains more prevalent in developed countries than in developing countries (Poushter, 2016).

Participatory media research tradition

Participatory media has a long history in the areas of communication for development and social change. This book acknowledges the work of activists in both the Global South and the Global North who have been at the forefront of participatory media engagement to effect political and social change. By theorising their practice these scholar/activists have shared their vital work with the global community and contributed to the growth of participatory media (Braden, 1999; Gumucio-Dagron, 2001; Johansson, 1999; Pink, 2001; Rennie, 2006; Riaño, 1994; Stuart, 1989; White, 2003; White and Patel, 1994). Within these volumes, one finds practice-led researchers – many of whom are women – whose contributions and commitment to community engagement laid down a solid foundation for participatory media practice and provided early navigational tools for those who have followed.

Their lessons have informed my own practice by engendering a culture of self-reflection, humility and above all, putting the community voice above the interests of the researcher. While these concepts are fluid and interact at various levels of meaning and purpose, they provide a novice participatory media user with a framework within which to discuss and study this important area in communication and media studies. Scholars who have studied participatory media's potential for social change are categorised below.

Participatory communication has its roots in the work of Latin American scholars such as Brazilian educator Paolo Freire (1970), Augusto Boal, Jesús Martín-Barbero and Luis Ramiro Beltrán. White et al. (1994) extended the dialogic/discursive approach in research and practice of participatory communication while Servaes (1999, 2001) and Carpentier (2009, 2016) have conducted ongoing scholarship in theory-building. Gumucio-Dagron (2001) & Gumucio-Dagron and Tufte (2006) bring fresh interpretations through their practice-based insights to the role of participatory media in cultural contexts. Tufte and Mefalopulos (2009), and more recently Hemer and Tufte (2016a), have made ongoing contributions to both revising and revisiting theories and concepts, as well as the practice of participatory communication. The field has evolved into Communication for Development and Social Change and is differentiated from Development Communication in their historical trajectories; the former arising from critical scholarship in Latin America which gave rise to participatory communication, and the latter from the World War II diffusion model that has progressively adopted participatory principles (Hemer and Tufte, 2016a).

Rodríguez (2001, 2011, 2016) has contributed to the field through her ground-breaking conceptualisation of participatory media as 'citizen's' media. She resists the term 'alternative' which, she argues, makes participatory communication alternative *to* something. Instead, she locates her research in Freire's philosophy of transformation and the feminist theory of radical democracy to theorise 'citizens' media as a process of personal empowerment for the human subject. She explains: "citizens' media is a concept that accounts for the processes of empowerment, conscientization, and fragmentation of power that result when men, women, and youth gain access to and re-claim their own media" (Rodríguez, 2004: 2).

New participatory culture

New media technologies and social media platforms have created significant transformations in the participatory culture. Henry Jenkins and his colleagues at Massachusetts Institute of Technology (MIT) provided one of the first insights into digital participatory media culture, which they explained as having "relatively low barriers to artistic expression and civic engagement, strong support for creating and sharing one's creation and some type of information mentorship whereby what is known by the most experienced is passed along to novices" (Jenkins et al., 2006: 3). They add that a participatory culture "is also one in which members believe their contributions matter, and feel some degree of social connection with one another" (Jenkins et al., 2006: 3).

Critical scholars question the participatory potential of Web 2.0 because it entrenches commercial and marketing exploitation of users in terms of data mining and corporate dominance, and promotes simplistic notions of participation which contribute to a self-branding culture, or worse still, fundamentalism and fascism (see Fuchs, 2017). It also creates an echo chamber where people gravitate towards people and information reflecting views which are similar to their own (Clark and Van Slyke, 2010). Fuchs has advanced the theory that while web 2.0 is a "computer-based networked system of human communication", the next level of digital technology development is web 3.0, "a computer-based networked system of human cooperation" (2017: 46). An example of this is Wikipedia which, he says, is "a new way of producing, owning, consuming and distributing goods and a new way of collaborative decision-making" (2017: 319). Delwiche and Henderson (2013: 4) contend that participatory culture studies can be thought of "as an emergent, interdisciplinary project". They provide a useful overview of studies in participatory culture by dividing these into four phases – 1985 to 1993 marked by active audience studies; 1994–1998 which saw popular use of the Internet and user friendly web browsers; 1999 to 2004 the arrival of web publishing systems and social media; and 2005–2011 which heralded the widespread use of broadband Internet connections and video-sharing site YouTube. To this we can now add the ubiquitous use of mobile technology. Unfortunately, Delwiche and Henderson overlook an important area of study by scholars of communication for development and social change.

A growing body of literature by contemporary scholars on media activism has engendered greater critical discourse in this sub-field of communication and media studies. Downing (2011) provided an early typology of alternative media as social movement media. Stephania Milan (2013: 2) looks at the role of communication technologies "as the digital backbone" in social justice struggles whereby "technology is not an end in itself; it is a means to a political end".

Practitioners

There are also growing numbers of media-based non-profit organisations that have contributed to the development of participatory media practice. InsightShare Network,[2] based in the United Kingdom, trains community filmmakers and participatory video facilitators in marginalised communities, both in developed and developing countries, to empower community action in response to local issues. Internews[3] is an international media NGO which supports media professionals and local activists in conflict and post-conflict zones, disaster situations, and in highlighting environmental, social and health issues. Civil society organisations are at the forefront of integrating participatory media tools for networking and activism. Food and Agriculture Organisation (FAO) has a long history of using participatory practices, including media, for education and knowledge sharing. UN agencies and the World Bank have integrated participatory media principles into their projects. However, there is debate as to whether participatory projects are truly participatory, as many NGOs use professional media producers to produce media in community-based projects. Participatory media use by vulnerable and marginalised communities will increase as these communities gain greater access to mobile networks and the means to share their stories.

Preview

Participatory media in environmental communication brings together selected stories of communities in the Pacific islands – a region that is severely affected by the impacts of climate change. The case studies and examples used in this book are from communities that have been at the margins of the digital revolution, but whose need to communicate remains critical because of a variety of environmental challenges. These perspectives have been largely absent in extant theorisation about participation and media in environmental communication.

Chapter two engages with interdisciplinary theories and concepts to arrive at a holistic communication model. Participatory Environmental Communication (PEC) weaves together three widely used concepts, *diversity, network and agency*, into a cohesive underlying system to bring scholars, practitioners and communities together in environmental communication.

Chapter three provides a pragmatic approach to engaging communities in participatory media by discussing the process of production. Attention is given to community engagement and the decision-making process in the project selection, the choice of equipment and message dissemination, together with research and ethical issues. Case studies provide insights into community use of different forms of participatory media for environmental communication.

Chapter four tells the story of the Pacific peoples and organisations who are at the frontline of impacts from climate change and natural disasters. Extreme weather events and storm surges in Fiji, Kiribati and Vanuatu have devastated the population. Despite that, citizens and governments strive to build resilient societies in the face of a changing reality.

Chapter five provides a background to the modern environmental movement, and examples of social networks. It then charts the emergence of the environmental movement in the Pacific region as a consequence of 50 years of nuclear testing and mining. This is followed by a discussion of contemporary concerns facing Pacific communities and the networks of solidarity that have resulted.

Chapter six explains the use of information and communication technologies (ICTs) in climate change communication in the Oceania region. It provides examples of different uses of ICTs for information dissemination, as well as local participation in communicative processes.

In the concluding Chapter seven the book urges scholars and practitioners to include participatory strategies in their toolbox of community engagement practices on environmental issues. It argues for increased adoption of participatory environmental communication as a way of ensuring greater diversity of voices at the local level.

Notes

1 www.fao.org/research-and-extension/en/
2 http://insightshare.org/network/
3 https://internews.org/

References

Andersen, P. 2007. *What Is Web 2.0? Ideas, Technologies and Implications for Education* [Online]. Bristol, UK: JISC Technology & Standards Watch. Retrieved from http://21stcenturywalton.pbworks.com/f/What%20is%20Web%202.0.pdf

Anderson, A. 2015. Reflections on environmental communication and the challenges of a new research agenda. *Environmental Communication*, 9(3): 379–383. doi:10.1080/17524032.2015.1044063

Atton, C. (ed.) 2015. *The Routledge Companion to Alternative and Community Media*. London: Routledge, Taylor and Francis Ltd.

Berndt, R. M. & Berndt, C. H. 1964. *The World of the First Australians*. Sydney: Lansdowne Press.

Boykoff, M. T. & Boykoff, J. M. 2007. Climate change and journalistic norms: A case-study of US mass-media coverage. *Geoforum*, 38(6): 1190–1204. doi:10.1016/j.geoforum.2007.01.008

Braden, S. 1999. Using video for research and representation: Basic human needs and critical pedagogy. *Journal of Educational Media*, 24(2): 117–129. doi:10.1080/1358165990240204

Brulle, R. J. 2010. From environmental campaigns to advancing the public dialog: Environmental communication for civic engagement. *Environmental Communication*, 4(1): 82–98. doi:10.1080/17524030903522397

Cabusao, A. B. 2012. Matigsalog women seed keepers: Bearers of life. *Mindanao Times*. Youtube.

Cameron, D. 2007. *Mobile Media and the Journalism Curriculum*. Mobile Media Conference, Sydney.

Carpentier, N. 2009. Participation is not enough: The conditions of possibility of mediated participatory practices. *European Journal of Communication*, 24(4): 407–420. doi:10.1177/0267323109345682

Carpentier, N. 2016. Beyond the ladder of participation: An analytical toolkit for the critical analysis of participatory media processes. *Javnost-The Public*, 23(1): 70–88. doi:10.1080/13183222.2016.1149760

Carvalho, A. 2010. Media (ted) discourses and climate change: A focus on political subjectivity and (dis) engagement. *Wiley Interdisciplinary Reviews: Climate Change*, 1(2): 172–179. doi:10.1002/wcc.13

Chi, H. & Lai, Y. 2015. The influence of 4G mobile value added services on usage intention: The mediating effect of perceived value of smartphone and phablet. *International Journal of Marketing Studies*, 7(6): 50. doi:10.5539/ijms.v7n6p50

Christian, D. 2011. *Maps of Time: An Introduction to Big History*. Berkeley, CA: University of California Press.

Clark, J. & Van Slyke, T. 2010. *Beyond the Echo Chamber: Reshaping Politics Through Networked Progressive Media*. New York: The New Press. Commoner, B. (1971). *The Closing Circle*. New York: Bantam Books.

Cox, R. 2007. Nature's "crisis disciplines": Does environmental communication have an ethical duty? *Environmental Communication*, 1(1): 5–20. doi:10.1080/17524030701333948

Cox, R. & Pezzullo, P. C. 2016. *Environmental Communication and the Public Sphere*. Thousand Oaks, CA: Sage Publications.

Curtin, M., Holt, J. & Sanson, K. (eds.) 2014. *Distribution Revolution: Conversations about the Digital Future of Film and Television*. Berkeley, CA: University of California Press.

Delwiche, A. & Henderson, J. J. (eds.) 2013. *The Participatory Cultures Handbook*. New York: Routledge.

Díaz Bordenave, J. 1989. *Participative Communication as a Part of Building the Participative Society*. A Key Concept in Communication for Change and Development, February 1989, University of Pune, India.

Downing, J. D. 2001. *Radical Media: Rebellious Communication and Social Movements*. Thousand Oaks, CA: Sage Publications.

Downing, J. D. (ed.) 2011. *Encyclopedia of Social Movement Media*. Thousand Oaks, CA: Sage Publications.

Falaki, A. & Adegbija, M. 2013. Investigating the use of the media in disseminating information on climate change in North Central Nigeria. *Global Media Journal-African Edition*, 7(1): 13–39. doi:10520/EJC139535

Farbotko, C. 2005. Tuvalu and climate change: Constructions of environmental displacement in the Sydney Morning Herald. *Geografiska Annaler: Series B, Human Geography*, 87(4): 279–293. doi:10.1111/j.0435-3684.2005.00199.x

Fenton, N. 2016. Alternative media and the question of power. *Journal of Alternative and Community Media*, 1: 10–11.

Flor, A. G. 2004. *Environmental Communication: Principles, Approaches, Strategies of Communication Appleid to Environmental Management*. Diliman, Quezon City: University of the Philippines, Open University.

Freire, P. 1970. *Pedagogy of the Oppressed*. New York: Herder and Herder.

Fuchs, C. 2017. *Social Media: A Critical Introduction*. London: Sage Publications.

Gumucio-Dagron, A. 2001. *Making Waves: Stories of Participatory Communication for Social Change*. New York: Rockefeller Foundation.

Gumucio-Dagron, A. 2014. Indigenous communication from multiculturalism to interculturality. *In:* Wilkins, K. G., Tufte, Thomas & Obregon, Rafael (eds.) *The Handbook of Development Communication and Social Change*. First ed. Chichester: John Wiley and Sons, Inc.

Gumucio-Dagron, A. & Tufte, T. 2006. *Communication for Social Change Anthology*. Orange, NJ: Communication for Social Change (CFSC) Consortium Inc.

Hansen, A. 2015. Promising directions for environmental communication research. *Environmental Communication*, 9(3): 384–391. doi:10.1080/17524032.2015.1044047

Harris, U. S. 2008. *Transforming Images: Participatory Video and Social Change in Fiji*. Doctor of Philosophy, Sydney: Macquarie University.

Harrison, T. M. & Barthel, B. 2009. Wielding new media in Web 2.0: Exploring the history of engagement with the collaborative construction of media products. *New Media & Society*, 11(1–2): 155–178. doi:10.1177/1461444808099580

Hemer, O. & Tufte, T. 2016a. Introduction: Why voice and matter. *In:* Hemer, O. & Tufte, T. (eds.) *Voice & Matter: Communication, Development and the Cultural Return*. Göteborg, Sweden: Nordicom, University of Gothenburg.

Hemer, O. & Tufte, T. (eds.) 2016b. *Voice & Matter: Communication, Development and the Cultural Return*. Göteborg, Sweden: Nordicom, University of Gothenburg.

Hulme, M. 2009. *Why We Disagree about Climate Change: Understanding Controversy, Inaction and Opportunity*. Cambridge: Cambridge University Press.

Jenkins, H. 2006. *Convergence Culture: Where Old and New Media Collide*. New York: New York University Pess.

Jenkins, H., Clinton, K., Purushotma, R., Robison, A. J. & Weigel, M. 2006. *Confronting the Challenges of Participatory Culture: Media Education for the 21st Century* [Online]. Chicago,IL: The MacArthur Foundation.

Retrieved from www.macfound.org/media/article_pdfs/JENKINS_WHITE_PAPER.PDF

Johansson, L. 1999. Participatory video and PRA: Acknowledging the politics of empowerment. *Forests, Trees and People Newsletter*, 40(41): 21–23.

Kheerajit, C. & Flor, A. G. 2013. Participatory development communication for natural resources management in Ratchaburi Province, Thailand. *Procedia – Social and Behavioral Sciences*, 103: 703–709. doi:10.1016/j.sbspro.2013.10.390

Lawlor, R. 1991. *Voices of the First Day: Awakening in the Aboriginal Dreamtime*. Rochester: Inner Traditions International.

Lie, R. & Mandler, A. 2009. *Video in Development: Filming for Rural Change*. Oxford, UK: CTA and FAO.

Lie, R. & Servaes, J. 2015. Disciplines in the field of communication for development and social change. *Communication Theory*, 25(2): 244–258. doi:10.1111/comt.12065

Lopez, V. M. (ed.) 2012. *Women's Wisdom: Documentation of Women's Knowledge in Agriculture (Case studies from Philippines, Thailand and Pakistan)*. Penang: Pesticide Action Network Asia and the Pacific.

Macchi, M. 2008. *Indigenous and Traditional Peoples and Climate Change* [Online]. Gland, Switzerland: IUCN. Retrieved from http://cmsdata.iucn.org/downloads/indigenous_peoples_climate_change.pdf

Meisner, M. 2015. *Environmental Communication: What It Is and Why It Matters* [Online]. Ohio: International Environmental Communication Network. Retrieved from https://theieca.org/sites/default/files/optp/%20OPTP%231-EC_What_and_Why.pdf

Melkote, S. R. & Steeves, H. L. 2001. *Communication for Development in the Third World: Theory and Practice for Empowerment*. London: Sage Publications.

Melkote, S. & Steeves, H. L. 2015. Place and role of development communication in directed social change: A review of the field. *Journal of Multicultural Discourses*, 10(3): 385–402. doi:10.1080/17447143.2015.1050030

Milan, S. 2013. *Social Movements and Their Technologies: Wiring Social Change*. New York: Palgrave Macmillan.

Milne, E-J., Mitchell, C. & de Lange, N. (eds.) 2012. *Handbook of Participatory Video*. Plymouth: AltaMira Press.Nanook of the North 1922. Documentary. Directed by Robert J. Flaherty. Retrieved from https://www.youtube.com/watch?v=m4kOIzMqso0

Nisbet, M. C. & Scheufele, D. A. 2009. What's next for science communication? Promising directions and lingering distractions. *American Journal of Botany*, 96(10): 1767–1778. doi:10.3732/ajb.0900041

NZ Herald. 2016. *Woman Live Tweets Cyclone Winston Destroying Her Home in Fiji*. 21 February. Retrieved from www.nzherald.co.nz/world/news/article.cfm?c_id=2&objectid=11593149

OCHA. 2016. *Koro Island: Tropical Cyclone Winston's Ground Zero* [Online]. United Nations Office for the Coordination of Humanitarian Affairs. Retrieved from 5 February 2018 www.unocha.org/story/koro-island-tropical-cyclone-winston%E2%80%99s-ground-zero

Olausson, U. 2011. "We're the ones to blame": Citizens' representations of climate change and the role of the media. *Environmental Communication: A Journal of Nature and Culture*, 5(3): 281–299. doi:10.1080/17524032.2011.585026

Pink, S. 2001. *Doing Visual Ethnography*. London: Sage Publications.

Poushter, J. 2016. *Smartphone Ownership and Internet Usage Continues to Climb in Emerging Economies: But Advanced Economies Still Have Higher Rates of Technology Use* [Online]. Washington, DC: Pew Research Center. Retrieved from www.diapoimansi.gr/PDF/pew_research%201.pdf

Rennie, E. 2006. *Community Media: A Global Introduction*. Lanham: Rowman & Littlefield Publishers Inc.

Rheingold, H. 2008a. Using participatory media and public voice to encourage civic engagement. *In:* Bennett, W. L., John, D. T. & Catherine, T. (eds.) *Civic Life Online: Learning How Digital Media Can Engage Youth*. Cambridge, MA: The MIT Press.

Rheingold, H. 2008b. Virtual communities-exchanging ideas through computer bulletin boards. *Journal of Virtual Worlds Research*, 1(1): 1–5.

Riaño, P. (ed.) 1994. *Women in Grassroots Communication: Furthering Social Change*. Thousands Oaks, CA: Sage Publications.

Rodríguez, C. 2001. *Fissures in the Mediascape: An International Study of Citizens' Media*. Cresskill, NJ: Hampton Press.

Rodríguez, C. 2004. The renaissance of citizens' media. *Media Development*, 2.

Rodríguez, C. 2011. *Citizens' Media against Armed Conflict: Disrupting Violence in Colombia*. Minneapolis: University of Minnesota Press.

Rodríguez, C. 2016. Human agency and media praxis: Re-centring alternative and community media research. *Journal of Alternative and Community Media*, 1: 36–38.

Santiaguel, A. F. 2010. The seed keepers' treasure. *Rice Today*. Retrieved from http://heirloomrice.com/pdfs/Seed_keepers_treasure.pdf

Schäfer, M. S., Ivanova, A. & Schmidt, A. 2012. *Media Attention for Climate Change around the World: Data from 27 Countries*. International Conference Culture, Politics, and Climate Change, 13–15 September 2012, The University of Colorado Boulder.

Servaes, J. 1999. *Communication for Development: One World, Multiple Cultures*. Cresskill, NJ: Hampton Press.

Servaes, J. 2001. Participatory communication (research) for social change: Old and new challenges. *Journal of International Communication*, 7(2): 5–13. doi:10.1080/13216597.2001.9751906

Spurgeon, C. 2015. Participatory media and "co-creative" storytelling. *Media International Australia*, 154: 132–138. doi:10.1177/1329878X1515400116

Stuart, S. 1989. Access to media: Placing video in the hands of the people. *Media Development*, 36(4): 8–11.

Tufte, T. & Mefalopulos, P. 2009. *Participatory Communication: A Practical Guide* [Online]. Washington, DC: The World Bank. Retrieved from http://siteresources.worldbank.org/EXTDEVCOMMENG/Resources/Participatorycommunication.pdf

Webster, B. 2017. Fijian storytelling can save world, says UN. *The Times*, 18 November. Retrieved from www.thetimes.co.uk/article/fijian-storytelling-can-save-world-says-un-gxx5t6zmr

White, S. A. (ed.) 2003. *Participatory Video: Images That Transform and Empower*. New Delhi: Sage Publications.

White, S. A., Nair, K. S. & Ascroft, J. 1994. *Participatory Communication: Working for Change and Development*. New Delhi: Sage Publications.

White, S. A. & Patel, P. K. 1994. Participatory message making with video: Revelations from studies in India and the USA. *In:* White, S. A., Nair, S. K. & Ascroft, J. (eds.) *Participatory Communication: Working for Change and Development*. New Delhi: Sage Publications.

Zhao, X. 2009. Media use and global warming perceptions: A snapshot of the reinforcing spirals. *Communication Research*, 36(5): 698–723. doi:10.1177/0093650209338911

2 Participatory environmental communication

A conceptual framework

Reflection

Saturday is market day in many Pacific island countries. I looked forward to market day in Nausori, the small river town where I grew up in Fiji. I can still recollect the colours, the sounds and the smells. It was the day our kitchen would be filled with the smell of fresh herbs, pineapples, mangoes and jackfruit. There was a sense of vibrancy in the town centre. People from surrounding districts would begin arriving at dawn as the sunrays of a new day touched the still waters of the wide Rewa River. Every form of transport was loaded with people and produce. Most arrived into town in buses and lorries, some in boats and a few in cars. Markets then and still remain the social, cultural and economic hub of Pacific island communities where exchange of goods, services and information take place. It is a place of cooperation, where a community of people collectively contribute to each other's livelihood and sustenance. Yes, people bargain for the best price and tempers flare in the tropical heat, but market vendors offer mutual support to each other, loading and unloading produce and minding each other's possessions. On Saturday nights, in those days, the market square became a place for dance parties for the youth. After a day of busy activity, the market arena was hosed clean and made ready for the next generation. The universal symbol of community markets offers us insight into how collaboration amongst diverse actors provides the foundation on which to build adaptable and resilient societies. Our earth is like a community marketplace, a biosphere, which supports a great diversity of life bound up in an intricate web of interrelation and interconnection while simultaneously maintaining the livelihood of individual species.

Introduction

Finding solutions to the world's worst environmental problems requires global cooperation at all levels of society. Unfortunately, a culture of denial, scepticism and apathy, aided by more than 50 years of strategic misinformation funded by vested interests in the fossil fuel industry (Klein, 2014; Foster, 2014), has stymied public action on urgent environmental challenges. The ongoing debate about the science of climate change is an example of an obstreperous communicative landscape where different meanings are derived from the same datasets and findings. Climate scientist and commentator, Mike Hulme, argues that "one of the reasons we disagree about climate change is because we believe different things about our duty to others, to Nature and to our deities" (2009: 144).

Generally, we are prepared to change our bad habits on doctor's advice at the onset of a life-threatening disease. Yet we fail to make the same connection between our individual action and its effect on the health of our planet. The baseline argument needs to shift from whether humans are responsible for global warming to one that asks, what is humankind's responsibility to our living planet? Every action we take must come with a reflection of the quality of life we choose to bequeath to future humans.

A participatory model of communication has an important role in disrupting centres of power in the face of draconian laws which inhibit citizen engagement in environmental protection. These information barriers are symptomatic of the powerbrokers who have shaped the meta-narrative about critical understanding of climate change that serve the political and economic imperatives of private capital and national development priorities. The mainstream news coverage presents these voices in binary opposition to the voices of people who demand environmental protection and stewardship.

The political and economic paradigms prioritise transnational corporate interests, affording them the rights of personhood at the same time as citizen's rights are being curtailed through the legal apparatus of government and international regimes. This can be seen in the form of increasing surveillance by government agencies, and promulgation of laws inhibiting protests. These actions violate civil liberties of citizens, as CIVICUS notes: "among the most targeted civil society activists are those who seek to uphold land, environment and indigenous people's rights" (CIVICUS, 2016: 73). It goes on to note: "Over 45% of all killing of human rights defenders were of environment, land and indigenous peoples' rights defenders". The worst attacks occurred in South and Latin America and South East Asia, with Brazil, Colombia, Honduras and the Philippines being the most dangerous countries for the activists (CIVICUS, 2016: 74).

Need for a holistic approach

How can science effectively communicate complex issues and at the same time inspire wider public engagement in the scientific sphere? What role can civil society play in the planet's renewal? How can our divergent ethical, moral and religious belief systems be brought into alignment to facilitate cohesive action for global environmental well-being? In his book *Storms of My Grandchildren*, climate change scientist James Hansen shares what changed him from being an objective observer of facts to an activist: "If it hadn't been for my grandchildren and my knowledge of what they face, I would have stayed focused on the pure science, and not persisted in pointing out its relevance to policy" (2010: xii). Pope Francis devoted his 2015 encyclical "On Care of Our Common Home" to environmental protection:

> I urgently appeal, then, for a new dialogue about how we are shaping the future of our planet. We need a conversation which includes everyone, since the environmental challenge we are undergoing, and its human roots, concern and affect us all.
>
> (2015: 5)

The Intergovernmental Panel on Climate Change (IPCC) also advocates an inclusive approach, stating: "Indigenous, local and traditional knowledge systems and practices, including indigenous peoples' holistic view of community and environment, are a major resource for adapting to climate change" (IPCC, 2014: 79–80).

Communication in this regard has to be transformative. It must move beyond top-down awareness-raising campaigns and invite people to participate in horizontal communication processes. If environmental communication is to be considered a 'crisis discipline', as Cox (2007: 5) suggests, then practitioners and scholars both should move beyond the sender-receiver transmission model towards a dialogic approach which will engage people in meaningful conversation. This change requires a shift from an individualistic win/lose mindset towards a collectivist approach.

In this chapter I present such a model. Participatory Environmental Communication (PEC) brings people together in a process of dialogue and collaboration which inspires collaborative action in the face of many environmental challenges. The Participatory Environmental Communication model is a process-oriented approach by which communities can share knowledge, create awareness and take action on environmental issues through their unique perspectives and intimate experiences of the environment in which they live.

The chapter advocates for a participatory model of communication that is able to meet the needs of diverse stakeholders as interrelated human actions on the environment converge to produce devastating results. A detailed discussion of the framework and concepts offers practitioners the ability to integrate diverse perspectives and networks in communication practice and resilience-building projects. The chapter provides an understanding of why a holistic approach is necessary for a better ecological future.

Towards a participatory model

As the impacts of climate change intensify to produce extreme weather events and natural disasters, there has been a notable overlap between environmental communication and risk communication. Risk communication has evolved from one-way dissemination model focused on behaviour change to two-way and multidirectional communication. It enables people at risk to take informed decisions to protect themselves and their loved ones. Risk communication utilises the most appropriate and trusted channels of communication. Communication techniques include the use of mass media, social media and community engagement. World Health Organisation notes "It requires a sound understanding of people's perceptions, concerns and beliefs as well as their knowledge and practices" including management of rumours and misinformation (WHO, 2018).

Scholars and practitioners have joined forces to search for best communication practices in disaster risk reduction and community resilience. The United Nations Office for Disaster Risk Reduction defines resilience as

> the ability of a system, community or society exposed to hazards to resist, absorb, accommodate, adapt to, transform and recover from the effects of a hazard in a timely and efficient manner,

> including through the preservation and restoration of its essential basic structures and functions through risk management.
>
> (UNISDR, 2018)

Within the context of Disaster Risk Reduction (DRR), capacity relates to the ability of people to overcome challenges or disasters by accessing and using resources which "include traditional/local knowledge, social organizations and solidarity networks, skills, and technologies" (Rey et al., 2017: 262). An example of this is found in the work of Janif et al. (2016) with 27 village communities in Fiji, where 'the apparatus of resilience remains in place'. They observe:

> Among the components of resilience are traditional environmental knowledge, community unity, respected decision-making processes, abundant communication and discussion opportunities, and community self-belief in their ability to manage their own futures.

There is increasing interest in the concept of resilience in the areas of global governance, national disaster policies and business and environmental movements. Critics point out that the concept of resilience has migrated from natural and physical sciences into the social sciences without due attention to the social relation and structures which are responsible for the disruptions (MacKinnon and Derickson, 2012). While top-down strategies place the burden of resilience and adaptation on communities in the face of mounting threats, capitalist social relations remain intact. Mackinnon and Derickson (2012: 255) suggest the use of "resourcefulness as an alternative concept to animate politics and activism that seek to transform social relations in more progressive, anti-capitalist and socially just ways". Notwithstanding the validity of this critique, in this book resilience is understood as the sum total of a community's ability to collectively recover using its own internal resources, as stated above in Janif et al. (2016).

The specific need to build resilience at the organisational level has exposed a gap in communication research. Rogers et al. (2016: 79) contend that there is a need for better understanding of how communication practices can facilitate collaborative engagement across organisations. They note: "Communication is a vital aspect of collaboration to build community resilience; however, communication researchers have yet to make significant inroads into the framing or implementation of building community resilience to disasters" (Rogers et al., 2016: 79). Reporting on a participatory action research

(PAR) workshop[1] which was organised to elicit how communication and resilience experts could collaborate to improve disaster prevention, management and mitigation practice, Rogers et al. (2016: 79) note:

> There is, however, no single model for "good" collaborative community engagement, no single set of tools that will foster healthy and sustainable communication practices in diverse local, regional, national and international situations and little evidence of a politically compatible inter-disciplinary approach that can build collaboration through action research.

If there is no single set of tools available for effective collaboration or to build community resilience to disasters, what communication resources can we use to develop a conceptual framework? How can communication scholars bring research and experiential knowledge from other fields of communication to fill the gap? To formulate a framework for Participatory Environmental Communication, I will draw on epistemic knowledge and practice-based insights from two areas – participatory communication and environmental communication.

At its core, participatory communication fosters dialogue, empowers people on the margins, gives voice to marginalised narratives, uses horizontal communication and engages an action-reflection-action cycle (Tufte and Mefalopulos, 2009). Participatory communication embeds local and indigenous knowledge because it is place-based and involves local communities in all phases of planning, research, project implementation and evaluation (White, 2003). These aspects are further explained in chapter 3 in the process of participatory media production. Meanwhile, environmental communication scholars seek to explain how communication shapes human-nature relations, influenced by various social, cultural, economic and political contexts in which these occur (Milstein, 2009: 345). Communication issues relate to climate change, environmental sustainability, biodiversity, sustainable development and representations of nature in popular culture. The interdisciplinary nature of the field demands that scholars bring in perspectives from multiple knowledge domains.

DNA of participatory environmental communication

The conceptual framework I propose has three interrelated elements – *Diversity, Network and Agency*. I refer to these as the *DNA* of participatory environmental communication, which uses systems thinking to invite whole-of-society participation.

These three elements are interconnected and integral to participatory environmental communication processes that create the condition for greater sustainability.

Diversity enables innovative and transformative thinking. The term here means both difference and inclusion of a broad range of factors – different knowledge systems, socio-cultural values and beliefs, abilities, talents, demographic variables (age, gender, class, ethnicity) and the non-human world (ecosystems, technologies, texts).

Networks are a complex system of relationships connecting both human and non-human worlds. They include human networks and non-human networks. The understanding that humans are but one part of this intricate web of creation, and not separate from it, is essential in the action-reflection-action cycle of the PEC process.

Agency is an action or a doing of human and more-than-human entities, which leads to an effect or outcome. Agency in the natural world contributes to the efficient working of an ecosystem. Human agency results from a realisation of our own potential through dialogic encounters that act as the catalyst for change.

It is useful to compare the *DNA* framework with a healthy ecosystem which is rich in biodiversity and capable of self-maintenance. Interaction and exchange are essential functions of a resilient ecosystem in which information from each component is fed back to maintain the system's stability. For example, human action in a rainforest can either nurture forest biodiversity or destroy the delicate balance essential for the survival of species. This interconnection and interrelation of the component parts is emphasised by American biologist Barry Commoner's First Law of Ecology: 'Everything Is Connected to Everything Else':

> Most ecosystems are so complex that the cycles are not simple circular paths, but are crisscrossed with branches to form a network or a fabric of interconnections. Like a net, in which each knot is connected to others by several strands, such a fabric can resist collapse better than a simple, unbranched circle of threads – which if cut anywhere breaks down as a whole. Environmental pollution is often a sign that ecological links have been cut and that the ecosystem has been artificially simplified and made more vulnerable to stress and to final collapse.
>
> (1971: 19)

Understanding how the ecosystem works provides a broad insight into the working of the social system in which communication is the dynamic force that gives meaning to social relations, as Harper (2012: 8) explains: "A sociocultural system is a network of interdependent actors (individuals, organizations, subsystems) that are in relatively stable patterns of interaction and intercommunication". With the rapid expansion of technology, a complex system of global interaction continues to expand, using diverse forms of transnational networks which impact on all actors.

One of the ways this interconnection can be explained is through systems thinking. It assists in our study and analysis of increasingly complex relationships in society. Systems thinking helps us to see things holistically because it highlights the interaction and interdependence of the elements, and the dynamic flows and feedback that assist in a system's success and maintenance. In developing a definition for systems thinking, Arnold and Wade (2015: 671) identified three things which are essential:

1. *Function, purpose, or goal:* This should describe the purpose of systems thinking in a way that can be clearly understood and relates to everyday life.
2. *Elements*: These elements will manifest as characteristics of systems thinking
3. *Interconnection*: This is the way the elements or characteristics feed into and relate to each other.

Using the three components identified above, I offer the following definition:

> The DNA framework advances collaboration and dialogue in society about environmental matters through diverse ways of knowing and being, using integrated networks to catalyse the agency of ordinary people towards collective action.

The DNA framework (Figure 2.1) identifies important attributes of each element – *diversity, network* and *agency* – which enable collaboration and dialogue among different actors, inviting them into a dynamic web of relationship. It is an integrated approach which can be replicated for different purposes, much like the DNA in living organisms, and should be integrated thoughtfully in participatory environmental communication processes. Each thread of the *DNA*

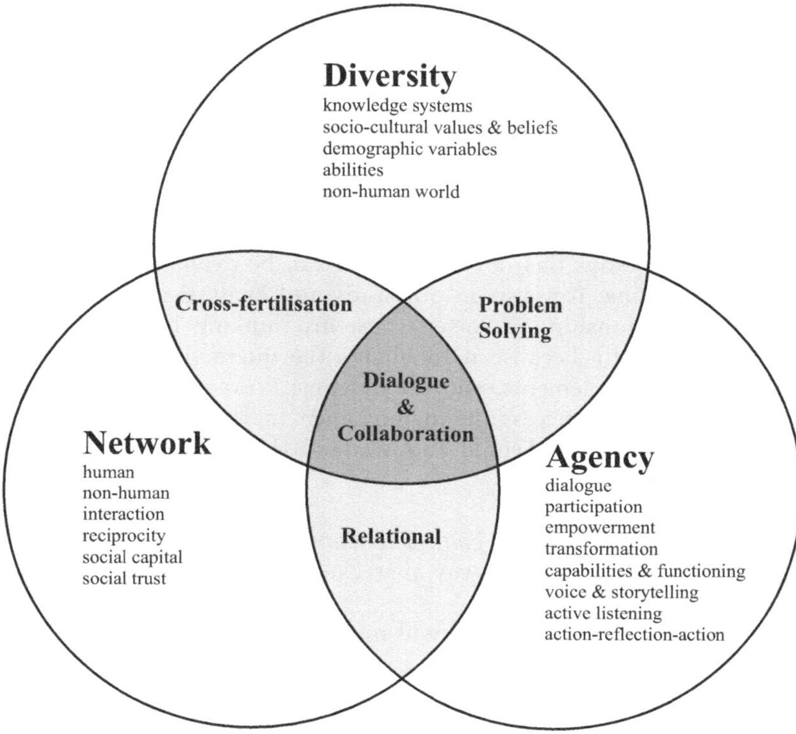

Figure 2.1 The DNA framework and its attributes

is not separate, but is intertwined in a holistic relationship, influencing and shaping the other in a communicative process, as illustrated below:

Diversity + network stimulates cross-fertilisation and cross-pollination of ideas by proactively linking disparate networks which previously may not have been linked or had weak links. These manifest as communication between different networks (as identified in Table 2.1) such as intergenerational, intercultural, interdisciplinary, interfaith and inter-agency dialogue, leading to innovation and collaboration previously untapped. The cross-sectoral partnership in Vanuatu in resilience-building is an example of this as discussed in Chapter 4.

Network + agency enable new relational possibilities. As actors in networks participate in reciprocal acts they begin to understand the

Table 2.1 DNA essentials

Elements	*Diversity*	*Network*	*Agency*
Definition	*Diversity* describes both difference and inclusion of a range of entities, abilities and ways of knowing.	*Networks* are complex systems of relationships and interactions (links) between human and non-human worlds (nodes) vital for exchange of information and resources.	*Agency* is the capacity of an actor to have the power to take action in a given context.
Attributes	knowledge systems socio-cultural values and beliefs demographic variables (age, gender, race, identities, education, place, disabilities etc.) interface non-human world biosphere holistic	human non-human relational (form and content) reciprocity social capital social trust social cohesion	dialogue participation empowerment transformation capabilities & functionings voice & storytelling active listening agents of change participatory action research
Manifestations	Communication cross-fertilisation new connection innovation intergenerational intercultural interdisciplinary interfaith inter-agency cross sector	Communication local or endogenous network heterogeneous network trusted network top-down horizontal bridging & bonding collaboration cooperation	Communication catalyst for change mutual learning problem-solving gather information, analyse, design, knowledge sharing transformative thinking awareness raising active participation marginalised voices action-reflection-action cycle

Source: Author

importance of their relational possibilities in expanding networks. The process of communication brings new insights into old problems. The relational aspects bring to light both positive and negative interactions. As such power, privilege and inequality should be studied alongside empowerment. This is demonstrated in two case studies – Wan Smolbag and Papuan Voices – in Chapters 3 and 6.

Agency + diversity facilitate problem-solving. Agency is a critical component in solving problems, both personal and societal. When people and organisations with different abilities, beliefs and knowledge systems interact proactively, transformational learning results. This is discussed in the case study of Andra Island Climate Change adaptation in Chapter 3.

Diversity + network + agency opens up a dynamic space for dialogue and collaboration. This is based on the belief that the cornerstone of human progress is collective enterprise instead of conflict and competition. The DNA evaluation of the participatory media process (see Table 3.4 in Chapter 3) illustrates the specific outcomes. A similar table can be created for other types of projects.

Comparison with other frameworks

There are numerous frameworks and tools available in the field of communication research. Two offer valuable comparisons to the DNA framework. Working within a communication for development (C4D) field, Lennie and Tacchi (2013) developed a comprehensive framework for evaluating local communicative ecology in communication for development initiatives. The international non-profit organisation, Internews, designed the Information Ecosystem for Resilience framework to represent the dynamic nature of information and communication flows during crisis (Internews, 2015a). Both these frameworks give priority to the context within which information and communication flows occur in different communities. Lennie and Tacchi (2013: 51) use a systems perspective to explain the holistic nature of their communicative ecology framework:

> The communicative ecology assumes that the role of a radio intervention or a facility such as a computer centre, for example, cannot be understood without understanding the wider communicative ecology of which it is a part, and the diverse positioning and communicative opportunities and experiences of people in any locality.

Communicative ecology mapping demonstrates gender, generational and class differences in communication use. It shows who has access to different information and communication channels and how factors such as geography, distance and cost can affect communication (Lennie and Tacchi, 2013).

Internews' Information Ecosystem framework to support resilience is defined as follows:

> Information ecosystems are complex adaptive systems that include information infrastructure, tools, media, producers, consumers, curators, and sharers. They are complex organizations of dynamic social relationships through which information moves and transforms in flows. Through information ecosystems, information appears as a master resource, like energy, the lack of which makes everything more difficult.
>
> (Internews, 2014: 12)

In the above definition, information is a critical component in building resilience because people need to exchange all types of information to adapt to their evolving environment. The Internews' Information Ecosystem framework includes eight dimensions that provide a holistic understanding and analysis of the information ecosystem of any given community or place. These include *Information needs* at the local level that are diverse and changing; *Information landscape* includes media, government, private industry and civil society; *Production and movement* maps different types of information flows available in a community; *Dynamic of access* includes barriers to interaction and participation affected by power dynamics; *Use of Information* looks at how it is processed, used and applied; *Impact of information* is the relationship between information, knowledge and behaviour change; *Social trust* or influence of trust networks on the flow and use of information; *Influencers* includes the people, organisations and institutions that shape information flows (Internews, 2015b: 14–17).

Both these approaches provide useful concepts and analytical resources that can be integrated in participatory environmental communication for mapping information and communication flows and for developing appropriate strategies for community inclusion.

However, the *DNA* framework is different from these other approaches because it actively seeks diversity in all forms, and emphasises the act of building networks at different levels – national, local and individual. It can be applied at the micro-level and scaled up depending on the context and need of different users. The *DNA* framework also places special emphasis on non-human networks, not limited to technology, but includes local ecosystems (inclusive of animals, plants, river systems etc.) as important nodes of communication. Their presence helps a community to observe and learn how their own interactions with these nodes impact on the resilience of the whole ecosystem. Participatory mapping of informational networks assists individuals and organisations to uncover the relational links (e.g. strong, weak, bonding, bridging), and the types of actors, resources and information that are available to them. Indigenous traditional knowledge is an important node in the wider network of knowledge holders.

As explained in chapter 1, the *DNA* framework repositions and reframes theories and concepts from other fields of scholarship to provide an alternative interpretation through the lens of participatory environmental communication. The *DNA* framework is both an analytical framework and a resilience-building tool designed to map information flows and identify access to resources and skills in multiple contexts. As such it is contextual and adaptable for a variety of uses by individuals, groups, organisations and communities. It can be employed in communication planning for both climate change adaptation and disaster risk management and also to monitor and evaluate participatory media projects, as seen in chapter 3.

The *DNA* essentials

This section discusses *diversity, network* and *agency* and their attributes, to illustrate how the dynamic flow of information, resources and skills through an integrated network offers new possibilities for broader citizen engagement in environmental communication. Each thread in the DNA manifests in different forms of communication and outcome. The attributes and manifestations for each element are identified in Table 2.1.

Diversity

Diversity is both rewarding and challenging. While it replenishes and offers wider choices, it also questions dominant patterns of relationships, worldviews and group privileges. More often than not, diversity makes a society more resilient because it endows rich sources of information, resources and skills. It fosters a wide range of perspectives, and brings competing and incompatible ideas, thus challenging established patterns of knowledge (Appiah, 2006). Diversity is inclusive of non-human worlds, scientific and traditional knowledge, sociocultural values and beliefs, demographic variables such as race, gender, age and (dis)abilities, to name a few.

Cultural diversity is an important concept in national and international policy formulations because it creates spaces for the cultural practices of minority communities. The United Nations Universal Declaration on Cultural Diversity[2] recognises cultural diversity as the common heritage of humanity:

> As a source of exchange, innovation and creativity, cultural diversity is as necessary for humankind as biodiversity is for nature. In this sense, it is the common heritage of humanity and should be recognized and affirmed for the benefit of present and future generations.
>
> (UNESCO, 2001)

Diversity has replaced the term equal opportunity in the workplace to mean inclusion of gender, minority groups and people with disabilities. It is increasingly favoured in the private sector where interdisciplinary research and partnership are valued resources for innovation and entrepreneurship, in contrast to silo thinking, which is considered inward-looking and limits cross-fertilisation of knowledge and access to resources. This is also true in academia where research and communication about climate change demands interdisciplinary approaches and integrative theories for problem-solving.

Participatory media, by its very nature, is diverse. Downing aptly describes it as a "media tapestry" which exhibits the following characteristics – "generally small-scale, in many different forms, that express an alternative vision to hegemonic policies, priorities, and perspectives" (Downing, 2001: v). To illustrate the multiplicity and richness of the field, Downing draws from a wide range of practices such as public speech, dance and songs, graffiti and dress, theatre, performance arts and jamming, the press, satirical prints, photomontage, posters and murals as well as the traditionally recognised media channels such as radio, film and television, Web 2.0 based Internet application and mobile technologies. This diversity of forms is its very strength, as the choice of participatory media is determined by the context and purpose.

Other ways of knowing

Disasters bring a rush of sympathy for human victims as the evening news anchor diligently enumerates the human lives lost, humans critically injured, homes incinerated or damaged and the cost to economy. But what about the animals caught in the fire? How many animals died, how many were injured, how many lost their homes, their babies and their sources of food? What would animals tell us, if we understood their languages, about their loss and damage following terrible wildfires in any part of the world? It is time media reports included non-human frames alongside human-interest frames in news reporting.

The non-human perspectives are missing from our anthropocentric worldview – one that only values the human condition. The modern human experience has been largely influenced by this dominant worldview, which has shaped the way in which we relate to Nature. Binary framework has been the focus of ecofeminist philosopher Val Plumwood's work, which sees the hyper-separation of humans from the natural world as a symptom of Western thought that has been responsible for other forms of dualism in the social world:

> In anthropocentric culture, nature and animals are constructed according to the same logic of the One and the Other, with nature as Other in relation to the human in much the same way that women are constructed as Other in relation to men, and those regarded as 'coloured' are constructed as Other in relation to those considered 'without colour', as 'white'.
>
> (Plumwood, 2002: 106)

Compare this with the principle of diversity in Buddhism, captured in this quote by Daisaku Ikeda, President of Soka Gakkai International (SGI), a Buddhist peace movement:

> The Buddhist worldview provides fertile ground for radiant diversity. For instance, Buddhism teaches that, without changing their nature, the cherry, the plum, the peach and the apricot all bloom in their own ways. None of them can or should be like the others. It is enough for each to shine in its own abundant individuality. Through analogy they teach us the right way to live.
>
> (SGI Quarterly, 2007: 6)

In comparing the two worldviews, Bai and Scutt (2009: 101) argue that the Western scientific worldview of dualistic modes of perception see mind (human) and matter (nature) as separate entities. The Buddhist practice of mindfulness, they proffer, is a "mode of consciousness that connects, integrates and transfuses everything", and through which the practitioner experiences 'the self-other continuity, unity and integrity' or 'interbeing'.

Herman (2016: 164) believes that the "disjuncture between Science, Culture and Traditional Knowledge in western thinking" is the cause of modern environmental problems: "Science gives the tools, but what we do with those tools is the product of culture. The wisdom to use these tools appropriately would have come from traditional knowledge". Based on his own field work in Hawaii with indigenous communities, Herman suggests that one way to bring culture, science and traditional knowledge into a single discourse is to think of the planet as an oceanic voyaging canoe used by Pacific islander seafarers who explored the vast ocean using their knowledge of the ocean and the skies in their double-hulled canoes (for example the Fijian *drua*). The success of the voyage depended on the following values; *Ike* (Knowledge), *po'okela* (the pursuit of excellence), *Kuleana* (responsibility and rights of individuals), *Pono* (to do the right thing) and *mãlama* (nurturance). He says to become good stewards of the earth, modern humans might benefit from thinking of themselves as voyaging in a canoe:

> all participants in it together, each needing to do his or her part, and all looking after and taking care of tasks, equipment, people, whatever, and with a common goal that is appropriate (*pono*) – then you will have a happier, more fulfilling, more successful and less harmful outcome.
>
> (Herman, 2016: 175)

Increasingly, science and academia recognise the importance of diverse forms of knowledge in meeting the challenges of a changing climate. Integration of traditional ecological knowledge (TEK) in resilience-building practices and climate change adaptation is actively sought to complement scientific knowledge (Khalil et al., 2016; McNamara et al., 2012; Ahsan and Azam, 2008). Pierotti and Wildcat remind us: "TEK is inherently multidisciplinary in that it links the human and the nonhuman, and is the basis not only for indigenous concepts of nature, but also for concepts of indigenous politics and ethics" (2000: 1333). As such, traditional ecological knowledge brings new insights and new concepts to Western knowledge systems and conventional adaptation practices. An example of this is found in chapter 4 in the case of the Regional Climate Traditional Knowledge project, whereby Pacific island communities and meteorologists have partnered to gather information about traditional forms of weather forecasts.

Network

Networks in natural and social systems enable the exchange of information and resources between two or more entities, species or groups. The network in nature is explained by Harper (2012: 3): "The most fundamental concept for ecological understanding is the notion of a *system* as a network of interconnected and interdependent parts". An ecosystem means the 'community' of things that live and interact in parts of the geophysical environment. Social networks are made up of nodes and relations:

> A network consists of nodes and links between those nodes. The nodes may be people, groups and organizations. The links may be social contacts, exchanges of information, political influence, money, joint membership in an organization, joint participation in specific events or many other aspects of human relationships.
>
> (Parks et al., 2005: 24)

Network analysis is a research method that is used to evaluate the relational links between actors (nodes) within a given boundary both horizontally and vertically, with special attention to power in the latter relationship (MacAulay, 2010). The relations between nodes can be considered in terms of form and content. "The form looks at the intensity of the relation as well as the level of joint participation of the actors" (MacAulay, 2010: 608). The form studies whether the connection between two actors is positive or negative, and what affects strong and weak ties. "The relational content of the connection between nodes considers the type of relations that is present" and the impact these have on both actors and the overall network (MacAulay, 2010: 608).

In the DNA framework, the relational links are the various manifestations of communication identified under *Network* and *Agency*. For example, stronger trust between networks would engender greater collaboration, resulting in opportunities for knowledge sharing and transformative thinking.

Human networks

Communication has been the cornerstone of human relationship and the glue that binds networks. It is a dynamic and dialogical process which connects members of a family, kinship group, community or organisation in society. Social capital, reciprocity and trust enable networks to function effectively.

A simple form of network is represented by a mother-child relationship. A newborn communicates with the mother by crying to alert her that it is feeding time and the mother provides the resource needed for the baby to survive and flourish. However, if that link is broken, then difficulties arise for both parties. If membership is limited to the same types of networks, the quality of communication and resources becomes restricted. For example, once the child grows up she will seek to expand her network beyond her parents to widen her opportunities for growth. Similarly, in the wider society diverse interconnections protect against vulnerabilities because different networks bring in new information, skills and resources.

Social capital

Social capital is an important attribute of social networks. Bourdieu (1986) defined social capital as the product of social obligations (connections) through membership in a group which provides each of its members with the backing of the collectively owned capital, a 'credential' which entitles them to credit, in the various senses of the word. Continued maintenance of social networks is crucial for this form of capital to accrue and for one to access it quickly.

The custom of looking out for each other in times of need is practised in many traditional societies. Fijian scholar Ropate Qalo has used the indigenous concept of *vakaviti* as forms of social capital (1998) "which enhance the common good more than the self-interest" (Huffer and Qalo, 2004: 108). These are:

Veilomani (genuine concern for one another)
Veikauwaitaki (caring and expressing concern for the wellbeing of others) *Veivakaliuci* (placing others ahead of oneself, supporting or looking up to them) *Veidokai* (honouring, respecting and upholding someone).

James Coleman gave the term social capital further structural definition by examining the various forms of "useful capital resources" that are constituted in social relations and social structures (Coleman, 2000). He identified three forms of social capital. The first is the obligations and expectations that people accumulate through reciprocal exchanges based on "trust that people have in each other". If I do you a favour, then I trust you to do me a favour sometime in the future. The second is the information-flow capability of social relations, whereby people seek out from their network new information, which can then become a basis for action. Participatory media provides one form of network for this kind of information flow, especially during disasters. Access to information is not the central reason for joining a social network but more a by-product of one's membership (Herreros, 2004: 14). The third form of social capital is the effective use of norms to both facilitate and constrain certain actions, e.g. rewarding selfless actions and disapproval of selfish actions. Trust is one of the key mechanisms that generates social capital. When one chooses to participate in a social network, it is generally based on a latent belief or a "trustful expectation" (Herreros, 2004: 9) that other members will also enter into a reciprocal arrangement with you. As such, trust is an important attribute in the maintenance of networks and flow of information, as seen in the identification of social trust in the Internews model discussed above.

Robert Putnam defined social capital as "connections among individuals – social networks and the norms of reciprocity and trustworthiness that arise from them" (2000: 19). He identified two important dimensions of social capital – bonding and bridging networks. While bonding networks "tend to reinforce exclusive identities and homogeneous groups" such as ethnic or gender-specific groups, bridging networks "are outward looking and encompass people across diverse social cleavages" and thus lean towards inclusiveness and "can generate broader identities and reciprocity" (Putnam, 2000: 22–23). The bridging dimension of social capital allows for the accommodation of difference while building on the element of trust to initiate dialogue between diverse groups, where communities can link with other communities across ethnic, gender, social and geographic divides. Strongly bonded groups are not problematic if they have people who have connections with other groups and can invite diversity. In this way it is not a single link but various links creating a 'mesh' effect (Leonard and Onyx, 2004: 185).

A number of academics debate the appropriateness of social capital as a concept (Haynes, 2009). Ben Fine cautions us about using the term social capital uncritically, especially as seen in its adoption by the World Bank and the development sector. According to him "social capital has a content and dynamic that severely constrain progressive developmental thinking" (2007: 573). Paul Haynes' (2009) critique is that the phrase social capital does not address the negative aspects, offers no sense of comparison and ignores the economics of a situation which other uses of the word 'capital' might demand.

The real problem lies in offering a more appropriate alternative to 'social capital'. There is no doubt that the term invites criticism by drawing on words which hold, in other contexts, different meanings. However, disregarding the specifics of terminology elsewhere, we need a phrase to adequately cover the concept of assets drawn from and offered back to community. So 'social' conveys that sense of community. We now require a term to indicate that the asset is of benefit to the community. The word 'capital' seems to serve that purpose adequately, because it implies something that adds to the storehouse of beneficial goods and services. So for the purposes of this book, 'social capital' means 'socially beneficial assets'.

In the context of the DNA framework, social capital is used to explain the form and content of various links between nodes. It is understood in the traditional cultural context as explained by Qalo (1998) as practised in kinship groups rather than the specific ways it may have been adopted in international development agenda. Any relationship comes with its inherent conflicts and power differentials and should be fully explored in discussions with various actors.

Reciprocity

Reciprocity generates unity and continuity. "Through reciprocity people can learn to trust. Consciously or unconsciously, as relationships develop, favours may be asked for or offered and returned. . . . It is about a give and take between equals" (Leonard and Onyx, 2004: 184). The act of giving and receiving is entrenched within the traditional value system of many indigenous societies. One example of reciprocity is found in Fiji, commonly referred to as *kerekere* or 'borrowing favour' and arises out of communal practice. As an important aspect of the Fijian social system, Ravuvu (1983: 250) presents the concept schematically in the following equation:

> Positive relationship = reciprocation = recognition, respect and appreciation = strengthening of the bond = incorporation and unity = continuity.
>
> Negative relationship = non-reciprocation = non-recognition, lack of respect and appreciation = weakening and breaking of the bond = separation and disunity = discontinuity.

Non-human networks

Actor Network Theory (ANT) is an influential cross-disciplinary approach that emerged in the 1980s from Science and Technologies Studies (STS) developed by French social theorists and STS scholars

Michel Callon and Bruno Latour, and British sociologist John Law (Callon, 1986; Law, 1992; Latour, 2005). In ANT culture, society and nature are created simultaneously (Latour, 2005: 281; Zammar, 2012) and are in a perpetual state of becoming. ANT considers both human and non-human elements equally as actors within a network and their identity is defined through their interaction with other actors. The terms actors/actants are used to describe the interaction between material and human factors in any process, an example being the interaction between human 'actors' and coastal ecosystems as 'actants', although some ANT scholars prefer to use actants for all things to diminish human and non-human dualism. The term "heterogeneous network" is used to overcome the "duality between human and non-humans" (Cressman, 2009: 4). Critics of ANT (Bloor, 1999), however, challenge its failure to acknowledge such basic social inequalities as race, class, gender and post-colonialism.

With digital affordances of Web 2.0, human networks have become enmeshed with non-human networks to multiply global connections. These forms of human/non-human networks include Internet-based social networks such as Twitter, Facebook, YouTube and WhatsApp, which contribute to our social and cultural identities and affiliations with the global community. These transnational communication networks enable the building of relationships and sharing of knowledge and information on a vast scale. In the Pacific context it has broadened the notion of kinship groups that are borderless, extending to other continents. With family links in New Zealand, Australia, the Unites States and Canada, the diaspora populations have become a main resource for island communities in times of crisis and natural disaster. Social media become a form of informal warning system about cyclones and tsunamis as relatives with better access to information send meteorological updates to those back home. These include twitter feeds of TV screens and screen shots of warnings from meteorological websites.

A growing area of study has broadened the concept of community by including non-humans as co-producers of knowledge. Researchers working on "more-than-human participatory research" treat these entities as "active stakeholders and collaborators in research and design" (More-Than-Human Lab, 2018). Studies include multispecies ethnography such as honey production by bees, animal and computer interaction and care for livestock (Reason, 2005; Birke, 2012).

In island studies, networks are an important framework of analysis, with island communities described as "simultaneously isolated and connected" (Papoutsaki and Harris, 2008; see also Baldacchino,

2006). Island networks enhance our understanding of interactions between humans and the natural environment because "Islands can be thought of as nodes of connectivity, as places in time where people, goods, plant and animal species, etc. come together" (Norder and Rijsdijk, 2016: 675). While islands are often referred to as places of vulnerability, they are also places of resilience by the nature of their isolation and bounded geography (Campbell, 2009). Island inhabitants have developed sophisticated knowledge systems and skills for self-preservation. These examples of island interconnectedness are further explored in Chapter 5 in the way they have been accessed by regional governments and civil society groups at international climate change negotiations. An example is the traditional water music in Vanuatu (see Case Study 2.1). It demonstrates the interconnectedness of humans and non-humans.

Case study 2.1 Women's water dance

On the island of Espiritu Santo in Vanuatu, women perform the mystical water music, produced by swirling and slapping the water surface to create a beautiful rhythmic tapestry of aquatic percussion. Their performance demonstrates the connection between humans and the hydrologic cycle that is the continuous movement of water around the surface of the earth. This story by Sandy Sur, manager of the Leweton Cultural Village, demonstrates how humans and more-than-human networks are connected and define each other's existence:

> The water music is played in the creek, it's played in the river and it is also played in the ocean. In different places it produces the sound of that nature. If it is the sea water, what comes out is the sound of the underwater from the ocean. When it is performed in the creek [it has the sound of] the running water . . . and the still water which sits there that we use to survive on. It gives the meaning of that area, how to remember, how to preserve the water, how important is the life of the water. A river or running stream gives us another rhythm, another sound, another connection between the water and the sound. The water and the ocean are connected; you can't divide that.

> The water music also connects the generations. We get the elders to do the performance with us, educating our next generation. The youth perform the music and also get the knowledge from the elders and understand what they're performing. As soon as you listen to the water, it is connecting you as human into your own nature. Water is the platform of connecting our nation and world itself.
>
> (S Sur, 2016, personal communication, 25 November)

Agency

Agency is an essential characteristic of resilience needed by adaptive systems to survive in a changing environment. Agency is the capacity of living things to act upon their environment (either social or physical) in order to bring about change. It is an empowering act by individuals or groups who have developed awareness of needs in their community. Hewson (2010: 13) notes: "Agency is the condition of activity rather than passivity. It refers to the experience of acting, doing things, making things happen, exerting power, being a subject of events, or controlling things". It is a process of conscientisation as well as a result of it. "Agency can be understood as the way in which people act on, or assert themselves in, their world . . . an element of self-determined action", according to Leonard and Onyx (2004: 23).

The attributes of agency identified in the DNA framework are considered through the interdisciplinary lens of two scholars who have given significant attention to subaltern perspectives. The culture-centred approach to social change by Mohan Dutta (2011) situates agency within structures that marginalise the subaltern sectors. The capabilities approach by the economist Amartya Sen, meanwhile, places individual freedom as both the end and the means of development.

Culture-centred approach

Dutta's conception of social change communication as a culture-centred approach is useful to consider in the context of agency in participatory environmental communication (Dutta, 2011). Dutta's culture-centred approach recognises subaltern agency as a starting point in enacting social change, thereby counterbalancing the hegemonic discourse which constructs the subaltern sectors as 'passive and without agency'. The culture-centred approach "concerns itself with the voices of marginalized groups and explores culture with structure that create conditions of marginality" (Dutta, 2011: 8). In developing a *culture-centred approach* to social change, Dutta emphasises the intersections between culture, structure and agency:

> Structure refers to the institutional roles, rules, practices and ways of organizing that constrain and enable access to resources. Culture constitutes the local contexts where meanings are continuously negotiated. Agency is the capacity of individuals and collectives to enact their choices as they negotiate structures.
>
> (2011: 40)

Dutta gives primacy to the voices of the 'subaltern' communities in "communicative processes to transform social structures", by creating participatory spaces which disrupt the hegemonic spaces of knowledge production. His framework is situated within subaltern studies and post-colonial theories (see Spivak, 1988; Said, 1978), both of which bring into question the centrality of western epistemology which "systematically erase ways of knowing that were native to the colonized spaces" (Dutta, 2011: 45). Dutta argues that the processes of globalisation within a neoliberal agenda have created "increasing inequalities locally and globally, with a limited number of people having access to the majority of the resources of the globe, as a large majority of others struggle to barely make a living" (2011: 49). Dutta's argument can be located within the environmental discourse in which the extractive industries have monopolised the wealth of communities while causing wholesale destruction of natural spaces that have little or no future potential for long-term sustainability of local communities (Dutta, 2011).

Capabilities approach

For Amartya Sen (1985: 203) agency is "what the person is free to do and achieve in pursuit of whatever goals or values he or she regards as important". Sen (2013) places human agency at the forefront of freedom and capabilities perspectives on transitioning to sustainability. He views the agency of humans as central to finding solutions to the environmental challenges which go beyond human needs and technical solutions to those of individual choice. Sen proposes that sustainability should be seen "in terms of sustaining human freedoms and capabilities" (2013: 1). In his view, the Brundtland commission's idea of sustainability, based on intergenerational justice, prioritises human needs rather than human freedom. The freedoms and capabilities approach speaks to such human pursuits as freedom to achieve dignity and respect as well as freedom from hunger, illiteracy, ill health etc. In proposing the freedom-based sustainable development, Sen gives as example the empowerment of women and the resulting decrease in fertility rates in India and Bangladesh. As women have achieved greater social freedoms in the forms of better education and increased employment, the fertility rate in Bangladesh has dropped from 6 to 3 children – without any coercion. Sen posits, "Behavioural change for fertility decline seems to work best with more freedom, not less"(2013: 6).

Agency is the person's ability to act on what they value and have reason to value. If people understand the impacts of environmental degradation on the things they value, then they would act to protect that which they value. These are the air we breathe, the soil that grows our food and the water that sustains our family. Sen's capabilities approach has three concepts which are useful in participatory environmental communication – resources, capabilities and functioning (Sen, 1993). In relating these to communication, resources are the modes and forms of communication, functioning (beings and doings) are access to these resources and capabilities are the actual skills in using technology to create the end product. However, an underlying concern of this approach is that it values individual above communal resources. In employing the capabilities approach it is important to keep in mind the all-powerful capitalist structures that permeate every facet of human endeavour in the current economic model (O'Hearn, 2009).

Applying capabilities approach in CDSC

Jacobson has suggested that the "capabilities approach to development should be applied as an overarching conceptual framework" by researchers working in the field of communication for development and social change (CDSC) because it offers a unified "framework for integrated analysis for many topics in development" (2016: 790). Both CDSC and the capabilities approach value the role of communication in development and are centred on citizen agency and voice. Jacobson identifies participatory communication along with social mobilisation and media advocacy, within the field of CDSC which address agency, an important concept in the capabilities approach. He states "Through its emphasis on public dialog, participatory communication shares an emphasis on citizen agency with the capabilities approach, often at the project level" (2016: 802). To make his point, Jacobson identifies participatory rural appraisal (PRA) and participatory learning action (PLA) approaches, developed in the 1970s and 1980s, which involved local people in the research process and later through PLA in project design, implementation and outcomes monitoring. Participatory media, especially video and radio, were used as participatory tools by communities to bring attention to their needs and in evaluation of projects. Jacobson argues that since communicative practices in participatory communication advance citizen agency "they comprise, in and of themselves, what Sen would treat as capabilities" (2016: 805). He identifies community radio as an area of research in CDSC, which could benefit from application of the capabilities approach framework (Jacobson, 2016).

Conclusion

Environmental communication is in need of a conceptual framework which includes the diversity of experiences, knowledge and practices of people and entities, other than the views of the society's elite. This chapter has presented a conceptual framework for participatory environmental communication which is inclusive, values diversity and engenders agency in building resilience in vulnerable communities. Three interrelated concepts of diversity, network and agency offer inclusive systems-based thinking which invites whole of society's participation in finding solutions to the current environmental crisis. The DNA framework aids our understanding of the rightful place of subaltern values and beliefs in matters of ecology and environmental disruption. The key to this interrelationship is to encourage dialogue and collaboration among diverse actors through alternative forms of communication. For example, mediated interpersonal communication or participatory media production, alongside face-to-face interaction, can be used to bring communities together as equal partners in a process which is sensitive to different viewpoints. When participatory environmental communication projects are successful they benefit all of society. These include better decision-making processes and improved interaction between members, which engender greater cohesion and trust within communities. They also build capacity for self-reliance and self-organisation. These elements are the essential building blocks of a sustainable and resilient society.

Notes

1 The workshop was hosted by the Centre for Climate Futures at Macquarie University (Australia) and was attended by this author
2 www.un-documents.net/udcd.htm

References

Ahsan, M. M. & Azam, M. 2008. Indigenous knowledge and cyclone preparedness of a remote coastal community in Bangladesh. *In:* Ahmed, M. F., Hossain, F. K. K. M., Khan, M. H., Jamil, S. & Matin, M. A. (eds.) *Indigenous People and Bangladesh Environment*. Dhaka, Bangladesh: Bangladesh Poribesh Andolon (BAPA) and Bangladesh Environment Network (BEN).

Appiah, K. A. 2006. Global citizenship. *Fordham Law Review*, 75: 2375–2391.

Arnold, R. D. & Wade, J. P. 2015. A definition of systems thinking: A systems approach. *Procedia Computer Science*, 44(2015): 669–678. doi.org/10.1016/j.procs.2015.03.050

Bai, H. & Scutt, G. 2009. Touching the earth with the heart of enlightened mind: The Buddhist practice of mindfulness for environmental education. *Canadian Journal of Environmental Education*, 14(1): 92–106.

Baldacchino, G. 2006. Islands, Island studies, Island studies journal. *Island Studies Journal*, 1(1): 3–18.

Birke, L. 2012. Unnamed others: How can thinking about "Animals" matter to feminist theorizing? *NORA-Nordic Journal of Feminist and Gender Research*, 20(2): 148–157. doi:10.1080/08038740.2012.674059

Bloor, D. 1999. Anti-Latour. *Studies in History and Philosophy of Science Part A*, 30(1): 81–112. doi:10.1016/S0039-3681(98)00038-7

Bourdieu, P. 1986. The forms of capital. *In:* Richardson, J. G. (ed.) *Handbook of Theory and Research for the Sociology of Education*. New York: Greenwood Press.

Callon, M. 1986. Some Elements of a sociology of translation: Domestication of the Scallops and the fishermen of St. Brieuc Bay. *In:* Law, J. (ed.) *Power, Action, and Belief: A New Sociology of Knowledge?* London: Routledge & Kegan Paul.

Campbell, J. 2009. Islandness: Vulnerability and resilience in Oceania. *Shima: The International Journal of Research into Island Cultures*, 3(1): 85–97.

Civicus. 2016. *State of Civil Society Report 2016* [Online]. Johannesburg: CIVICUS World Alliance for Citizen Participation. Retrieved from www.civicus.org/images/documents/SOCS2016/summaries/SoCS-full-review.pdf

Coleman, J. S. 2000. Social capital in the creation of human capital. *In:* Dasgupta, P. & Serageldin, I. (eds.) *Social Capital: A Multifaceted Perspective*. Washington, DC: The World Bank.

Commoner, B. 1971. *The Closing Circle*. New York: Bantam Books.

Cox, R. 2007. Nature's "crisis disciplines": Does environmental communication have an ethical duty? *Environmental Communication*, 1(1): 5–20. doi:10.1080/17524030701333948

Cressman, D. 2009. *A Brief Overview of Actor-Network Theory: Punctualization, Heterogeneous Engineering & Translation*. British Columbia, Canada. Simon Fraser University ACT Lab/Centre for Policy Research on Science & Technology (CPROST), 1–17.

Downing, J. D. 2001. *Radical Media: Rebellious Communication and Social Movements*. Thousand Oaks, CA: Sage Publications.

Dutta, M. J. 2011. *Communicating Social Change: Structure, Culture, and Agency*. New York: Routledge.

Fine, B. 2007. Social capital. *Development in Practice*, 17(4/5): 566–574.

Foster, J. 2014. *After Sustainability: Denial, Hope, Retrieval*. Oxon: Routledge.

Francis, P. 2015. Encyclical Letter Laudato Si' of the Holy Father Francis On care for our common home. Retrieved from http://w2.vatican.va/content/dam/francesco/pdf/encyclicals/documents/papa-francesco_20150524_enciclica-laudato-si_en.pdf

Hansen, J. 2010. *Storms of My Grandchildren: The Truth about the Coming Climate Catastrophe and Our Last Chance to Save Humanity*. New York: Bloomsbury Publishing.

Harper, C. L. 2012. *Environment and Society: Human Perspectives on Environmental Issues*. Boston, MA: Prentice Hall.

Haynes, P. 2009. *Before Going any Further with Social Capital: Eight Key Criticisms to Address* [Online]. Valencia, Spain: Institute of Innovation and Knowledge Management, INGENIO (CSIC-UPV) Retrieved from http://digital.csic.es/bitstream/10261/14203/1/Before_Going_Any_Further_With_Social_Capital__Eight_Key_Criticisms_to_Address[1].pdf

Herman, R. D. K. 2016. Traditional knowledge in a time of crisis: Climate change, culture and communication. *Sustainability Science*, 11(1): 163–176. doi:10.1007/s11625-015-0305-9

Herreros, F. 2004. *The Problem of Forming Social Capital: Why Trust?* New York: Palgrave Macmillan.

Hewson, M. 2010. Agency. *In:* Mills, A. J., Durepos, G. & Wiebe, E. (eds.) *Encyclopedia of Case Study Research*. Thousand Oaks, CA: Sage Publications.

Huffer, E. & Qalo, R. 2004. Have we been thinking upside-down? The contemporary emergence of Pacific theoretical thought. *The Contemporary Pacific*, 16(1): 87–116. doi:10.1353/cp.2004.0011

Hulme, M. 2009. *Why We Disagree about Climate Change: Understanding Controversy, Inaction and Opportunity*. Cambridge: Cambridge University Press.

Internews. 2014. *Information Ecosystems in Action: New York* [Online]. Arcata, CA: Internews. Retrieved from www.internews.org/sites/default/files/resources/Internews_Information_Ecosystems_in_Action-NewYork.pdf

Internews. 2015a. *Mapping Information Ecosystems to Support Resilience* [Online]. Arcata, CA. Retrieved from www.internews.org/sites/default/files/resources/Internews_Mapping_Information_Ecosystems_2015.pdf

Internews. 2015b. *Why Information Matters: A Foundation for Resilience* [Online]. Arcata, CA: Internews. Retrieved from www.internews.org/sites/default/files/resources/150513-Internews_WhyInformationMatters.pdf

IPCC. 2014. *Climate Change 2014: Synthesis Report* [Online]. Geneva, Switzerland: The Intergovernmental Panel on Climate Change. Retrieved from http://ipcc.ch/report/ar5/syr/

Jacobson, T. L. 2016. Amartya Sen's capabilities approach and communication for development and social change. *Journal of Communication*, 66(5): 789–810. 10.1111/jcom.12252

Janif, S. Z., Nunn, P. D., Geraghty, P., Aalbersberg, W., Thomas, F. R. & Camailakeba, M. 2016. Value of traditional oral narratives in building climate-change resilience: Insights from rural communities in Fiji. *Ecology and Society*, 21(2): 7. doi:10.5751/ ES-08100-210207

Khalil, M. B., Jacobs, B. C. & Kuruppu, N. 2016. Grassroots technologies and community trust in climate change adaptation: Learning from Coastal settlements of Bangladesh. *In:* Filho, W. L. (ed.) *Innovation in Climate Change Adaptation*. Switzerland: Springer.

Klein, N. 2014. *This Changes Everything: Capitalism vs. the Climate*. New York: Simon & Schuster.

Latour, B. 2005. *Reassembling the Social: An Introduction to Actor-Network Theory*. New York: Oxford University Press.

Law, J. 1992. Notes on the theory of actor-network: Ordering, strategy and heterogeneity. *Systems Practice*, 5: 379–393.

Lennie, J. & Tacchi, J. 2013. *Evaluating Communication for Development: A Framework for Social Change*. Oxon: Routledge.

Leonard, R. & Onyx, J. 2004. *Social Capital and Community Building: Spinning Straw into Gold*. London: Janus Publishing.

MacAulay, K. D. 2010. Network analysis. *In:* Mills, A. J., Durepos, G. & Wiebe, E. (eds.) *Encyclopedia of Case Study Research*. Thousand Oaks, CA: Sage Publications.

MacKinnon, D. & Derickson, K. D. 2012. From resilience to resourcefulness: A critique of resilience policy and activism. *Progress in Human Geography*, 37(2): 253–270. doi:10.1177/0309132512454775

McNamara, K. E., Hemstock, S. L. & Holland, E. A. 2012. *PACE-SD Guidebook: Participatory Vulnerability and Adaptation Assessment* [Online]. Suva, Fiji: Pacific Centre for Environment and Sustainable Development, The University of the South Pacific. Retrieved from http://eugcca.usp.ac.fj/Portals/0/PACE-GB.pdf

Milstein, T. (ed.) 2009. *Environmental Communication Theories*. Thousand Oaks, CA: Sage Publications.

More-Than-Human Lab. 2018. *What We Do* [Online]. Aotearoa, New Zealand: Victoria University of Wellington. Retrieved 9 January 2018 from http://morethanhumanlab.org/what-we-do/

Norder, S. J. & Rijsdijk, K. F. 2016. Interdisciplinary island studies: Connecting the social sciences, natural sciences and humanities. *Island Studies Journal*, 11(2): 673–686.

O'Hearn, D. 2009. Amartya Sen's development as freedom: Ten years later. *Policy & Practice: A Development Education Review*, 8: 9–15.

Papoutsaki, E. & Harris, U. S. (eds.) 2008. *Unpacking "Islandness" in South Pacific Islands Communication*. Singapore: Asian Media Information and Communication Centre.

Parks, W., Grey-Felder, D., Hunt, J. & Byrne, A. 2005. *Who Measures Change? An Introduction to Participatory Monitoring and Evaluation of Communication for Social Change*. Orange, NJ: Communication for Social Change (CFSC) Consortium Inc.

Pierotti, R. & Wildcat, D. 2000. Traditional ecological knowledge: The third alternative (commentary). *Ecological Applications*, 10(5): 1333–1340.

Plumwood, V. 2002. *Environmental Culture: The Ecological Crisis of Reason*. London: Routledge.

Putnam, R. D. 2000. *Bowling Alone: The Collapse and Revival of American Community*. New York: Simon & Schuster.

Qalo, R. R. 1998. *Vaka-Viti: If It Is Social Capital, Then We May Be More Advanced than We Think*. Oceania Centre for Arts and Culture Seminar Series, 14 October, 1998, The University of the South Pacific, Suva, Fiji.

Ravuvu, A. 1983. *Vaka i Taukei: The Fijian Way of Life*. Suva, Fiji: Institute of Pacific Studies, The University of the South Pacific.

Reason, P. 2005. Living as part of the whole: The implications of participation. *Journal of Curriculum and Pedagogy*, 2(2): 35–41. doi:10.1080/15505170.2005.10411540

Rey, T., Le De, L., Leone, F. & Gilbert, D. 2017. An integrative approach to understand vulnerability and resilience post-disaster: The 2015 cyclone Pam in urban Vanuatu as case study. *Disaster Prevention and Management: An International Journal*, 26(3): 259–275. doi:10.1108/DPM-07-2016-0137

Rogers, P., Burnside-Lawry, J., Dragisic, J. & Mills, C. 2016. Collaboration and communication: Building a research agenda and way of working towards community disaster resilience. *Disaster Prevention and Management*, 25(1): 75–90. doi:10.1108/DPM-01-2015-0013

Said, E. 1978. *Orientalism: Western representations of the Orient*. New York: Pantheon.

Sen, A. 1985. Well-being, agency and freedom: The Dewey lectures 1984. *The Journal of Philosophy*, 82(4): 169–221. doi:10.2307/2026184

Sen, A. 1993. Capability and well-being. *In:* Nussbaum, M. & Sen, A. (eds.) *The Quality of Life*. Oxford: Oxford University Press.

Sen, A. 2013. The ends and means of sustainability. *Journal of Human Development and Capabilities*, 14(1): 6–20. doi:10.1080/19452829.2012.747492

SGI Quarterly. 2007. Talanoa-talking from the heart: Interview with Dr. Sitiveni Halapua. *SGI Quarterly*: 47. Retrieved from www.sgi.org/content/files/resources/sgi-quarterly-magazine/0701_47.pdf

Spivak, G. C. 1988. Can the subaltern speak? *In:* Morris, R. C. (ed.) *Can the Subaltern Speak? Reflections on the History of an Idea*. New York: Columbia University Press.

Tufte, T. & Mefalopulos, P. 2009. *Participatory Communication: A Practical Guide* [Online]. Washington, DC: The World Bank. Retrieved from http://siteresources.worldbank.org/EXTDEVCOMMENG/Resources/Participatorycommunication.pdf

UNESCO. 2001. *Universal Declaration on Cultural Diversity* [Online]. Paris: United Nations Educational, Scientific and Cultural Organization. Retrieved 25 February 2018 from www.un-documents.net/udcd.htm

UNISDR. 2018. *Terminology* [Online]. Geneva, Switzerland: United Nations Office for Disaster Risk Reduction. Retrieved 19 April 2018 from www.unisdr.org/we/inform/terminology

White, S. A. (ed.) 2003. *Participatory Video: Images that Transform and Empower*. New Delhi: Sage Publications.

WHO. 2018. *Risk Communication: General Information on Risk Communication* [Online]. Geneva, Switzerland: World Health Organization. Retrieved from www.who.int/risk-communication/background/en/

Zammar, N. 2012. Social network sites: The science of building and maintaining online communities, a perspective from actor-network theory. *In:* Tatnall, A. (ed.) *Social Influences on Information and Communication Technology Innovations*. Hershey, PA: IGI Global.

3 Participatory media

Pedagogy and practice

Reflection

The women organised themselves around the table and sat waiting to view the first pictures of their shoot on the television screen. An expectant silence descended in the room. Their eyes were focused towards the screen. They watched with quiet pride at their own images. Some laughed out loudly, some commenting on the composition of the picture. The women who were being interviewed seemed at ease because they were being interviewed by people they knew and the interviews were in their own language. When technology is used by ordinary people, it conforms to their ways of life. The camera came into their life space. The interviews took place in their homes, the community halls – familiar places in the women's lives. Location became an important aspect of their representation. The subjects of discussion were their lives, their skills and their communities, at which they were experts, and they spoke about these things with great ease. It was important for these rural women to record their contribution to the community and the camera gave them the means. In the process the women had proved that technology was no barrier to their storytelling (Harris, 2008).

Introduction

For the first time in human history we have communication technologies and platforms that enable all peoples of this world to have a voice. Interestingly, the people who have the most power also make the most use of information and communication technologies. We now know, through research into these communities, that it is not that disenfranchised people don't have anything to say, rather that they have lacked access to the enabling technologies to make their voices heard. Public participation increases when communities have access to specific tools and the knowledge to use them.

Participatory environmental communication (PEC) is designed to give prominence to local issues through a process of engagement which engenders dialogue and collaboration across diverse communities. The bottom-up model of communication gives voice to those on the margins including women, indigenous groups and people with disabilities.

Technological affordances allow previously silent communities to tell their own stories with easy-to-use audio-visual devices, mobile phones and online sharing platforms. By telling their own stories in their own way, people are able to communicate compelling accounts of how different, but interrelated, environmental, political and economic issues converge and impact at a local level. People take the initiative to observe changes, understand what it means at the local level, and identify appropriate solutions for the future (Faris et al., 2014). These local stories, told through personal narratives of adversity and triumph, bring a lived human dimension to environmental changes occurring on a global scale.

This chapter provides an understanding of participatory media pedagogy and practice – the undergirding principles, the tools, the methods – and its potential within the field of environmental communication. It demonstrates the power of participatory media as a dialogical tool by including case studies of the ways communities have used media to raise awareness of, and find solutions to, environmental challenges. The final section includes a step-by-step explanation of the participatory production process – choice of project, the process of message creation, and how engagement in the participatory process invites diversity, strengthens networks and fosters agency in participants. Product dissemination and important issues around consent and evaluation are highlighted. Examples of case studies provide further insight into choice of media and process. These projects demonstrate how participatory engagement shifts the debate away from alarmist rhetoric and towards an action-reflection-action cycle through which communities learn about food and water security, resources management, jobs and livelihoods.

Participation and permutation

Participation in communication activities is a basic human right enshrined in Article 19 of the Universal Declaration of Human Rights:

"Everyone has the right to freedom of opinion and expression; this right includes freedom to hold opinions without interference and to seek, receive and impart information and ideas through any media and regardless of frontiers". This move to democratise communication at international, national and local levels has enabled access and participation by the grassroots in communication systems.

Communication scholars have argued for a model where participation is seen as an end in itself rather than as a means to an end (Diaz-Bordenave, 1989; Melkote and Steeves, 2001; White et al., 1994). According to Servaes (1999), the focus shifts from creating a need for the information being disseminated to one of disseminating information for which there is a need. Melkote and Steeves (2001: 336) have identified communitarian theory along with post-structuralism and postmodernism as the "assumptive basis for participatory strategies".

In 1969, Sherry Arnstein created a 'ladder of citizen participation' to demonstrate how powerholders manipulate participatory processes (Arnstein, 1969). In her eight rungs of participation, the bottom rungs signify non-participation, the middle rungs are akin to tokenism and the top rungs represent citizen power. For Arnstein, "citizen participation is a categorical term for citizen power" (1969: 216). Taking Arnstein's model further, Carpentier (2016) has developed a four-level and 12-step analytical model to build a theory of participation within the political approach to unravel the complexity of participatory processes. To bring greater insight into the complex nature of participation, his model focuses on the process, field, actor, decision-making moment and power.

Participation is generally thought of as healthy in a society. However, not every act of participation necessarily guarantees a good outcome. People can participate in violent acts, gossip, socially and politically regressive movements and acts of domination that disempower individuals and society as a whole.

Barriers to participation

Everyone has a voice, but there are obstacles to listening to voices of the marginalised. Participation can be used to gain power, not only over one's own life, but over the lives of others. Barriers to participation can come in many forms. Social and cultural practices or communal hierarchy suppress the voices of the marginalised, such as women and youth. Sometimes barriers are created by the very people who are in positions of power to help marginalised voices, as illustrated in this example by community radio trainer and researcher Kalinga Seneviratne:

> We were in Bangladesh conducting a UNESCO funded community media training with marginalised communities. We wanted to produce a program with people working in construction zones. The Bangladeshi officials who were supporting the training told us that if you go there people won't talk to you because they are mainly women who are too busy working on the construction site. They emphasised, they will not talk to you. We went there anyway and found women were carrying buckets of sand on their head walking up the hill to the construction site. When one woman saw us with the recorder, she threw the bucket on the ground, grabbed the microphone and started berating the construction site owner about how they have been exploited.
>
> (K Seneviratne, 2016, personal communication, 21 January)

These self-appointed benefactors who speak for marginalised groups are usually reluctant to transfer power and skills to the people they purport to represent. In my own experience of community media, the technocrats who appoint themselves in training positions or those in managerial roles stage a power struggle when minority groups are ready to take on the roles. This leads to conflicts and fragmenting of community organisations and resources.

In a review of 100 case studies, Gaventa and Barrett (2012) created a typology of four democratic and developmental outcomes of citizen engagement: construction of citizenship, practices of citizen participation, responsive and accountable states, and inclusive and cohesive societies. While their study found an overall positive effect from citizen participation, they noted that bureaucratic inertia to citizen engagement was one of the biggest barriers, alongside political reprisals, against those who dared to challenge institutional power (Gaventa and Barrett, 2012).

Civic activities fail to include the poorest and most marginalised in society, according to Mansuri and Rao (2013). Instead, the profile that emerges from review of literature is that "participants in civic activities tend to be wealthier, more educated, of higher social status, male and more politically connected than nonparticipants" (Mansuri and Rao, 2013: 5).

The most educated members involved in participatory projects tend to hold more influential positions in those projects, either actively self-nominating themselves or by default in the absence of a group protocol. When people communicate from a disempowered position, their voices become lost or misrepresented by those who mediate on their behalf.

Participatory methodology can duplicate the social hierarchy at the grassroots. This may happen through the acts of exclusion, self-exclusion and non-participation (Kothari, 2002). The most marginalised may not participate through self-exclusion. Employing Foucault's analysis that "power is everywhere . . . and is thus found in the creation of norms and social and cultural practices at all levels", Kothari cautions practitioners that participatory techniques may actually lead to

> reassertion of power and social control not only by certain individuals and groups, but also of particular bodies of knowledge. What counts as 'local knowledge' is very often the effect of specific kinds of techniques of power, of regulation and of normalization.
> (Kothari, 2002: 141, 142)

Participation as empowerment

Paulo Freire's seminal writings and his earlier work with impoverished people in Brazil are largely credited for the emergence of the empowerment framework. He developed a method of teaching for illiterates and the oppressed, published in 1970 as *Pedagogy of the Oppressed*. Students have to be actively involved, through praxis and dialogue, for genuine learning to take place. Freire differentiated this from banking education where one person acts on another (Freire, 1970). By gaining an ability to 'name the world' through a process of dialogue, human beings are able to free themselves from the broader ills or oppression.

Social action researchers who have facilitated participatory media (PM) activities demonstrate that agency, voice, empowerment and transformation are more than mere words on a page when media spaces are created for marginalised groups who have been silent in the past. These words become living representations of people who provide concrete meaning of these concepts through action. In a study of PV for disaster and climate change risk reduction with women and children in Nepal, Plush (2009) found that their viewpoints shed a new understanding on recurrent problems. Women and children were able to present alternative solutions to community-based adaptation programmes, which differed from outside interventions driven by experts. Plush observes:

> As climate scientists and decision-makers reach to the micro-level to deepen their knowledge on community impacts, participatory

> video with marginalized groups has the potential to not only provide data, but offer prioritized solutions in the voice of those most impacted.
>
> (2009: 26)

In a similar study designed to increase young people's knowledge of disaster and climate change issues in the Philippines, Haynes and Tanner (2015: 367) note that the films made by youth identified not only the impacts but also the 'underlying social causes of disaster'. Through these skills, vulnerable communities are better able to communicate and represent their interest in national policy development and international forums on climate change.

Climate change practitioners in the Pacific favour a participatory approach to awareness-raising and a move away from the 'one-way communication paradigm'. Noting that participatory approaches were 'extremely effective', they identified activities to invite community participation: the development of seasonal calendars, communal storytelling, poem and song competitions, and the use of books and movies (McNamara, 2013: 868).

The role of critical participatory media has added benefits other than citizens' rights to communicate, as Spurgeon (2015: 133) argues: "By devoting energy and resources to extending storytelling capacity on a population-wide basis, societies may also be developing tools to explore the health of communal life and thereby become more robust and adaptable".

Participatory media engagement in message production becomes a transformative experience as Pileggi and Morgan (2017: 161) explain:

> However contradictory this may seem, as the storyteller transcends boundaries through telling the story, she opens possibilities of experiencing other ways of being in the world. This transformative quality of storytelling is a critical element in environmental communication, which also strives to transcend boundaries.

Case study 3.1 Wan Smolbag community theatre

> "Everywhere, acting is what people like to do"
>
> – *Joanne Dorras, Founder of Wan Smolbag.*

- *Project:* The Turtle Theatre Project was designed to raise awareness about the excessive killing of turtles and consumption of turtle eggs in Vanuatu. Through participatory theatre, Wan Smolbag encouraged the Ni-Vanuatu community to understand why turtles are important for the environment and people. Instead of killing turtles, the theatre project hoped to persuade communities to change their behaviour and protect turtle nesting grounds.
- *Context:* Wan Smolbag is a community theatre group in Vanuatu that produces plays on a range of social and environmental topics. It was founded by Joanne Dorras and Peter Walker, who moved to Vanuatu from England 26 years ago. "Song and harmony are so much a part of this culture, so it is relatively easy to produce plays because you could write songs and people would make up a tune or harmonise" says Dorras, who writes the plays. In addition to the main travelling theatre, there are several amateur groups for youth and people with disabilities. Based in Port Vila, Wan Smolbag also offers support to disadvantaged youth through health, nutrition, agriculture, arts and craft, music and sports programmes.
- *Process:* Before writing the play, researchers interviewed communities on Efate, the main island of Vanuatu, to incorporate local community perspectives and find traditional stories about turtles. The play *I am a Turtle* is a story told by a turtle about to be killed. It weaves in traditional stories to explain the ecological impact of killing turtles. The theatre group travelled to different villages in North Efate to perform the play. At the end of the play, actors invited community members to discuss the issue. Dorras confirms the dialogic aspects of participatory theatre: "What I find most amazing is how much communities will discuss. Over the years I've seen a lot of dialogue about all sorts of different issues – domestic violence, sanitation in the village, conservation, family planning".
- *Outcome:* The new awareness has led to some villages placing a tabu on the killing of turtles and the eating of turtle eggs.

A Turtle Monitor network was set up with community involvement to watch for nesting turtles, tag turtles and collect data on hatchlings and adult turtles. The network has over 400 monitors based in coastal villages on different islands in Vanuatu.

The project

Communication for Development (C4D) practitioners use diverse forms of communication to facilitate participation. These include face-to-face meetings; ICTs such as community radio, participatory video (PV), digital storytelling; entertainment education including television and radio dramas, theatre and music; mobile phones and social media (Lennie and Tacchi, 2013). Kalas and Finlay (2009: 16) recognise that ICTs such as community radio, mobile phones and Internet-based applications "facilitate networking and building of coalitions" while they simultaneously encourage partnerships between different stakeholders.

PEC activities enable citizens to gather information about their local fauna and flora or share knowledge about adaptation methods, land management, sustainable farming, etc. The process of production helps participants to understand immediate risks by recording changes to their environment and to learn about the impacts and adaptation methods by talking to environmental experts. Participatory approaches have been used successfully for conflict resolution between communities over water, land and mining disputes.

Communities and individuals empowered with appropriate skills want to have ownership of their own storytelling instead of having outsiders tell their stories. Outsiders are often surprised to find how well people are able to articulate issues and concerns in their community, and how insightful and creative they are in finding sustainable solutions. This belies common assumptions held by project planners that communities are disempowered and lack education or understanding about issues affecting them. People do not need lessons on what to say. Instead, they need support on how best to make themselves heard.

Whose knowledge?

Both participants and project planners should establish what reciprocal arrangements exist between parties for the exchange of knowledge and resources. Has the community developed the idea? Has someone else approached the community? It is important to recognise the "tensions, contradictions, dilemmas, and power imbalances inherent in all forms of knowledge production and communication" when "scientists and citizens engage in mutual learning on the basis of the different knowledge forms that they bring with them" (Phillips et al., 2012: 4; Phillips, 2011). Antonella Piccolella observes that a mutual distrust exists between scientists and local communities in relation to each other's knowledge. She used participatory three-dimensional modelling as a mapping process for climate change adaptation in Boe Boe, Solomon Islands and advocates bridging the divide through participatory means:

> The necessary integration of such diverse worldviews can be possible if adequate tools and methodologies are available for multi-stakeholder dialogue. These tools would allow scientific data to be verified against local data, and provide a medium for local communities to communicate credibly with policy-makers.
>
> (2013: 29)

Donna Haraway's (1988) notion of *situated knowledge* provides a compelling argument to value knowledge which is "based on people's lives, the location and the specific" (Liao, 2006: 104). For Haraway knowledge can't be totally objective because all types of knowledge are influenced by the variables of time, space, values and beliefs, among other factors.

Who will participate?

Establishing trust with the community is vital to the success of participatory projects. Finding a community leader who has the trust of all sections of the community and the authority to engage with them is essential. This may include the local social worker, faith-based leader or president of a social club.

Norris (2002: 3) states that "rich and dense associational networks facilitate the underlying conditions for interpersonal trust, tolerance and cooperation". Uslaner and Conley (2003: 333, 335) distinguish between the types of trust based on people's engagement in either outward-looking or inward-looking groups. Outward-looking people are *generalised trusters*. They are willing to trust strangers, and believe that sharing common values and social interaction with people unlike themselves can be rewarding. They are the ones most likely to form bridging social capital. Inward-looking people are *particularised trusters*, who may play an active role within their own social groups but are less likely to participate in "civic engagement in the larger community".

Leonard and Onyx posit that isolated communities do not need to 'shift' from bonding to bridging in order to 'get ahead', but may find other ways to forge links with other communities, such as seeking the assistance of a 'trusted professional' who may become a valuable ambassador in this process. In the PV project I facilitated with rural women in Fiji, I found three types of groups – the indigenous Fijian women, the long-term Indo-Fijian settlers and the newly arrived Indo-Fijian displaced farmers. The indigenous women and the long-term settlers enjoyed greater trust with each other and were enthusiastic participants in the PV project, while the new arrivals were reluctant participants. Their participation was a direct result of their trust in the local social worker who invited them to the workshop (Harris, 2008).

Diversity mapping tool

A diversity mapping exercise at an early stage will ensure that a cross-section of the community is included. The diversity mapping creates a visual representation of the forms of diversity one would find in different contexts. These can be expanded further to uncover other related links as the context changes. Creating a diversity matrix for a community, no matter how big or small, assists both community members and outsiders to understand the complexity and information needs, especially during disasters and for resilience-building.

The map would also identify information exchanges specific to different groups in a society made up of diverse ethnocultural groupings. Unequal distribution of resources is often a cause of conflict. Neglected communities perceive aid disparity during emergencies as favouritism for one group over another or racial discrimination, instead of a possible oversight on the part of agencies who are unaware of the appropriate channels through which information is exchanged in specific groups.

Table 3.1 provides an example of a diversity matrix which can be considered when creating resilience-building tools. By undertaking a diversity mapping exercise, people are forced to consider their actions on the environment, including connections with non-human entities such as river systems, forests, oceans and animals that have either direct or indirect impact on a community's livelihood. Participants would identify resources, skills and knowledge already present

Table 3.1 Diversity matrix

Knowledge systems	*Cultural factors*	*Social factors*	*Non-human*	*Modes of communication*	*Services*
Scientific	Values/beliefs	Education/ literacy	River systems	Interpersonal	Emergency/DRR
Indigenous/ Traditional	Race/ethnic	Age	Coastal ecosystems	Print	Medical
	Kinship/clan				
Local	Place-based	Gender/ LGBTI	Ocean	Broadcast	Community
Experiential	City/rural	Disabilities/ abilities	Animal population	Internet/mobile	Private/commercial
Faith-based	Religion		Plant varieties	Transport	Faith-based
Ecological	Language	Wealth	Soil biome	Non-human	Ecosystem

Source: Author

within the group and those they would have to source from outside. For example, the community may have a strong base in local knowledge, but require current scientific knowledge to help them understand soil salinity or reasons for a decrease in soil quality that is affecting their crop output.

In designing disaster prevention and management plans, each element in the column is identified and mapped according to the local context. The elements can be deleted or new ones added, based on different contexts and changing needs. Once the diversity matrix has been created, community members undertake a network mapping exercise to identify the relational links between nodes and their value in either strengthening or weakening individual, family or community resilience in a particular context. Once identified, communities can then engage constructively in the participatory mapping process of vulnerability assessment and adaptation planning. Diversity mapping initiates a culture of integrated problem-solving and a move away from thinking in silos. The identification of these actors and entities is the first step towards building networks.

Choosing the right fit

Decisions about the forms of participatory activity to use should be made in consultation with the community. Young people may prefer to use video and social media, while older members may prefer radio or traditional forms of face-to-face communication. Consider the audience and final dissemination of the end product. Will it be distributed globally or shared for local information only?

An aim of the DNA framework is to reveal connections between people and organisations and how well these connections enable information flows, i.e. how, where and with whom people exchange information and the quality of these exchanges within a given context. Table 3.2 gives examples of participatory communication flows in a community setting.

These below-the-surface information flows are also vital in the early hours and days of emergency response and aid distribution. The intrinsic information exchanges may be quite obvious to individuals in the local community, but not so obvious to outsiders such as first responders and aid workers when they reach the community. Community participants and aid agencies can use this tool to identify the community information flows in the local area.

Table 3.2 Participatory communication flows

Face to face	*Community hubs*	*Internet*	*Mobile*	*Broadcast*	*Print*	*Traditional*
Family	Civic spaces	Websites	Smartphones	Community radio	Community	Conch shells
Neighbours	Marketplace	Social media	MMS	Community TV	Newspapers	Chalk drawing
Friends	Community hall	Emails	SMS	Fixed-line phones	Magazines	Sand drawing
Chiefs	Parks, Plaza	Blogs	Phone apps	HF/shortwave radio	Flyers	Traditional drums
Faith-based leaders	Faith-based places	Wikis	Messaging apps	Ham radio	Posters	Bells
Teachers	Schools	Soundcloud			Brochures	Animal
Government	Social clubs				Letters	Pigeons
Service providers					Newsletters	Plant behaviour
NGO					Notice boards	

Source: Author

Equipment

Once the decision is made about the form of media, a list is drawn up based on the available budget. Diverse forms of equipment allow greater participation of people with different abilities. For example, vision-impaired people in the community are able to use digital audio recorders to record stories and songs while digital photography may be more suitable for younger children. However, issues of sustainability must be factored in when using sophisticated digital equipment. These include maintenance and repair in remote communities and training of new members if assigned facilitators leave the community.

If using digital media, smaller and lighter cameras are recommended, due to ease of use and storage. Consumer model cameras provide good picture quality and digital storage options, simple editing features and easy playback. Some cameras incorporate a built-in projector to allow communities who don't have easy access to electricity to review the footage using battery power. Good quality sound is achieved with an external microphone or by changing settings of the in-built microphone. It is important that a sturdy weather-proof case is provided to store all equipment and accessories. Also essential is an equipment checklist that records all equipment loaned, the names of borrowers, and all items returned. Here's an example of a list of equipment we used for PV training after consultation with the participants:

1. A consumer model camera with built-in projector to review the footage by projecting it on any plain surface instead of through a TV monitor. Other features included a built-in GPS for monitoring climate change impacts in an area, and hybrid storage with in-camera flash memory and an external SD card.
2. A waterproof digital still camera for underwater photography and during rain events.
3. A digital sound recorder for audio-only recording of interviews, songs and sound effects to produce podcasts and radio programmes.
4. Waterproof and shockproof carrying case for easy transport and storage.
5. MacBook Pro laptop with iMovie editing software.
6. Accessories such as tripod, headphones, cable extension and batteries.

At the time of purchase in 2012, iMovie provided intuitive and easy to operate software that came with the computer. Beware of buying editing software which is too sophisticated and demands many hours of training.

Mobile technologies provide low budget and easy to use alternatives. Smartphones are widely accessible even in remote communities, and provide the ability to record, edit and distribute from the palm of one's hand. In recent years, I have used android smartphones and a free video-editing app to train NGO workers and youth. The apps provide easy to use editing features such as titles, transitions and the ability to record and insert voiceover and music. The end product is easily shared on social media with specific users or for wider distribution.

My research in the Pacific Islands has revealed that younger participants want access to media platforms which enable them to record and share content instantly (Figure 3.1). In training community artists and NGO workers, I found that both groups had access to smartphones with audio-visual recording capability, but neither group had proactively used these features for environmental communication. Knowledge of the mobile phone as a multimedia tool generated great excitement

Figure 3.1 Kiribati Climate Action Network members use smartphones at a participatory media workshop.

Source: Usha S. Harris

among participants, as it was more immediate and accessible than the process of shooting and editing stories using video cameras. Once they understood the theory and practice of participatory media, participants were enthusiastic about its use in their work. They wanted to capture interviews about community concerns and solutions while on field visits, and create better workflow by producing reports in audiovisual form. While adept at using smartphones, participants said learning to use an editing app (such as Filmora Go, a free android video editor) had filled an important knowledge gap which made it possible for them to instantly record, edit and share content. The editing app was a midway solution, providing a more polished content with titles, voiceover and music that would have greater appeal in the wider community. Notwithstanding the aesthetics of producing content which could be considered amateur, there remains the question of what is more important to the viewer – relevance of the information being shared or the quality of content, or both.

Marcus Williams, a media educator who participated in the Pacific Mama's project in New Zealand, makes an interesting observation about the quality of the finished PV product:

> What I discovered through this workshop is that participatory video has a mechanism that will self-regulate this problem, which might be seen as a bit of a paradox. The reflective dimension of the process ensures that the protagonists see the quality and any deficit issues, for themselves. This then creates the motivation for technical improvement in them. It takes time and you have to be patient and allow that learning to occur rather than just to come in and take control, but ultimately, the objective will be achieved because media is now a shared language and we are all a great deal more expert than we realise. By acknowledging this through the participatory video methodology, far greater insights into the subjects, community or issue at hand are possible.
>
> (Saifoloi et al., 2016: 12)

Case study 3.2 Andra Island climate change adaptation[1]

- *Project:* In a study conducted with the Andra Island community of Manus Province in Papua New Guinea, researcher Aaron Inamara used photography and storytelling to incorporate indigenous knowledge (IK) into community-based adaptation (CBA) strategies. A creative media workshop with 13 participants from Andra Island was facilitated to examine the issue of climate change adaptation. Several photo essays were created.

- *Context:* Inamara, who is from Madang Province in PNG, involved both the local indigenous and provincial institutions in negotiating the participation of the community. Endorsements by community leaders created an atmosphere of trust that motivated people to engage in the research. This research added a new dimension to the work being carried out by the Wildlife Conservation Society (WCS), by building basic creative media capacity to promote local adaptive strategies on Andra Island. Creative media was seen as an incentive to the revival of Andra Island's indigenous knowledge.
- *Process:* The visual participatory action research process encouraged the community to explore the strengths and weaknesses of their social and cultural capitals, providing insights into alternative measures for climate change adaptation. The visual and action research design and methods were incorporated within a Melanesian indigenous knowledge context. The participants produced photographic essays recording indigenous knowledge in food production and preservation to increase Andra Islanders' capacity to respond to food insecurity. Projects focused on oral tradition and intergenerational knowledge transfer in the use of traditional fishing nets; maintaining traditional water holes; benefits of a papai fish trap; clamshell farming as a traditional mariculture; line fishing in the face of declining fish stock; the socio-economic value of barter trade between Andra Islanders and mainlanders; and the preservation of sago, a staple crop.
- *Outcome:* A photo exhibition was held within the community after the completion of the workshop. Through photo essays, members of the community were able to reflect on adaptation practices involving indigenous knowledge. The islanders were keen to develop the concept of photographic essay production as they felt it had great potential in reviving useful cultural practices to help address negative climate change impacts.

The process

This section provides an explanation of the participatory production process – breaking down the production as a transactional dialogic process. Communities' active involvement, or agency, is essential to the production process, as opposed to the product-driven media in which communities may feature but not participate. The participatory action research (PAR) approach of ongoing dialogue and action-reflection cycle enables participants to engage in a dynamic process of building relationships where "communication is truly something we do *with* others and not *to* them" (Adler and Rodman, 2003: 28).

Dialogue is an important first stage of project planning in participatory environmental communication whereby the information needs of the community are identified. Dialogue engenders trust and agency. The essential first step in finding effective solutions is understanding people's concerns and aspirations. It is useful to explore with the community the assumptions and understanding of their own capacity to make a difference. Discussions can explore these questions:

1. What do we already know about the issue (awareness of the issue)?
2. What do we not know? What do we know that others don't know (gaps in knowledge)?
3. What action do we need to take to gain new knowledge (agency)?
4. What is our intrinsic motivation for action on this issue (problem solving)?

    a. Does it matter to me (salience)?
    b. Moral responsibility to address this issue (motivation)
    c. Will my action make a difference (efficacy)?
    d. Will other citizens join me/will I receive organisational support (capacity for collective action)?

(informed by Johari window in Tufte and Mefalopulos, 2009)

Project design

A purpose for the project should be established with community input, which in turn will influence the design of the PEC project. In the project design, consider the various production phases and specific factors which may impact the process. The workshop schedule is designed to be flexible in order to accommodate participants' own needs and different skill levels. The workshop is open to a range of participants, including people with disabilities and those lacking in skills or literacy; it is open to the ideas, voices and needs of the community; it is not scripted and adopts production values which reflect the norms of the community. The production process is shown in table 3.3.

This section will focus on participatory video (PV) process. There are five phases in the PV production process – camera training, story development, location shoots, viewing sessions and editing. Participants are encouraged to experience every facet of the production process with progressive transfer of knowledge and skills from the facilitator to the group. The essence of the narrative and the end product must reflect the decision-making process of the people themselves.

The first day of training is designed as a media literacy session to increase the participants' knowledge of media. It elicits responses about participants' own negative and positive experiences of media and why media access is important for lay people. This follows a basic lesson in visual grammar, such as photo composition and shot sizes. The participants should gain access to the tools (smartphones, cameras and audio recorders) on the first day of training to maintain their interest and ensure their continued participation in the workshop. The first exercise is to teach participants to hold the camera and take one

Table 3.3 The production process

Pre-production	1. Contact with the community (trust-building). 2. Decision by individuals to participate in the project (self-efficacy). 3. Becoming part of the team (network and collaboration). 4. Focusing the project (purpose or objective).
Production	5. Sketching the idea into *our* story (finding the voice). 6. Training and rehearsal sessions. Interactions with technology (overcoming the mental and social blocks). 7. Decisions as to the roles of producer, director, actor, technical operators (finding one's capability). a. How is she/he chosen? b. Does this follow the social hierarchical order? c. How does it affect message creation? d. What is the group dynamic? 8. What and who to include (aesthetics and voice).
Post-production	9. Viewing and Feedback: Self-reflection and changing self-perception. 10. Editing: What remains and what is left out? Who decides? 11. Screening the finished project and community discussion.

Source: Author

well composed shot. In the second exercise participants are paired; one speaks directly to the camera sharing something about herself, e.g. her background, her strengths, her views on the topic covered, while being recorded by another participant. As participants gain confidence and skill using the camera, other exercises can be conducted, such as two-person and group interviews, allowing them to use a tripod and external microphones.

The editing phase is where the participatory elements can become eroded. Lunch and Lunch (2006) remind us that editing done by facilitators or donors can make the participatory video process tokenistic, because the final message may turn out to be completely different from that which the participants intended. An important part of the participatory decision-making process is the ongoing cycle that involves screening back and group discussion on what should be included and excluded from the final product. Drawing a paper storyboard is another useful method to get more people involved and to reach consensus on the type of editing required, even without the availability of a computer.

Facilitator's role

There is a high degree of collaboration between facilitators and participants during the production process. Facilitators must have gained participatory media training themselves so that they are able to constructively assist communities during group workshops and production activities. Good communication skills between the facilitator and various groups in the community are required in order to interpret diverse needs, resolve conflicts and ensure equitable use of equipment. It is necessary for facilitators to create a road map and structure the workshop (Shaw and Robertson, 1997) in a way that would allow a good idea to be integrated, reflecting Galbraith's (1991) observation that the transactional process requires "collaboration, support, respect, freedom, equality, critical reflection, critical analysis, challenge, and praxis" from both facilitator and learner. Facilitators can influence the success or failure of the PV process through their own knowledge and experience of media production, knowledge of community, enthusiasm and ability to incorporate an egalitarian approach in the facilitation of the workshop.

A crucial element of the participatory process is self-reflection. One has to be constantly self-reflective. How am I participating in the process? There is no right or wrong answer. Facilitators must learn to stop controlling the moment. Let the subject control the moment because this is a collective process of discovery.

Giving up control can be difficult for first-time participatory media facilitators, especially those who have worked in mainstream media. They approach participatory workshops with a professional mindset where visual aesthetics and the end product are more important than the process of engagement. This struggle is shared by Sandra Kailahi, a television broadcast journalist who participated in a PV workshop:

> In the case of broadcast media, the journalist will use a camera-person to film the story, then edit the piece without any direct contribution from the community other than those being interviewed. There is a chain of editorial command before the final story goes to air and at no time is the community consulted over what actually goes to air. This is the complete opposite to the participatory video approach, which involves the community from the beginning to the end and it is indeed the group that decides on the final product, as they are involved with every step of the process.
>
> (Saifoloi et al., 2016: 14)

As a participatory media facilitator, I have worked with migrant and refugee groups, rural women, youth groups, postgraduate students, NGO workers and academics. Every participatory media experience is different because every community is different. Some will be extremely rewarding and some will be difficult. If communities are brought together without proper consultation it is difficult to create a truly collaborative experience. Lack of trust, disengagement with the project and cultural taboos about speaking publicly also affect participation. Difficulties arise when the participants and the subject matter are chosen by the funding organisation. I have worked on projects where people were brought together in a forced way and they didn't really want to be there. Some of my most enthusiastic participants have been rural women in Pacific island communities who have self-nominated their projects.

Conflict is another major area of concern in participatory projects. Chin points to the difficulty experienced by practitioners who had "sufficient facilitation and participatory communication skills but totally untrained in conflict management and resolution" (2000: 31). The readjustment of power relationships, disagreement on priorities, goals and methods of doing things, as well as larger underlying social issues facing the communities, may emerge as the causes of conflict, he notes.

Example of participatory video workshop

The following example discusses an eight-day train-the-trainer participatory video workshop on gender and climate change which I conducted with community facilitators from the Cook Islands, Fiji, Guam, Kiribati, Niue and Papua New Guinea. As professionals working in science, law, media and non-government organisations, they were already disseminating information about climate change and were adept at using ICTs.

The participants were well informed about climate change issues in the Pacific and brought a mix of perspectives to enable informed discussions, followed by hands-on training with video and audio equipment and collaborative group work. There were four sessions per day, each having a particular focus on building media literacy and technological skills. Participants actively engaged in production and post-production exercises while they developed climate change themes for short video stories. The central focus of all activity was to enable participants to become confident in the use of video and audio equipment and editing software so that they were able to invite community engagement in content creation within their own communities.

On the first day participants were introduced to some of the key terms and concepts. They were specifically asked to consider what the term 'participation' meant to them; to consider some of its challenges; to identify their own methods of working with the community; and to determine how they may be able to integrate participatory approaches to their work on climate change.

On the second day the group was asked to collaboratively identify key themes/topics on climate change in their region. They then divided into teams of two to develop story ideas based on these themes. Intensive training in all aspects of video production skills and digital editing continued throughout the week. For their final project, each participant identified a community with whom they could produce a ten-minute video story on climate change impacts or adaptation. During the group viewing sessions, participants exchanged feedback on each other's work as an important part of the action and reflection cycle, where participants compared common threads and divergences in problem-solving (see Harris, 2014). All the projects were edited by the participants themselves. The completed projects were shared online on Vimeo and YouTube.[2]

These projects demonstrate how participatory video can raise awareness about the impacts of climate change on coastal communities, and become a tool for knowledge sharing. Two of the participants produced instructional videos as a teaching tool to improve communities' income generation and livelihood capacities. One team recorded the step-by-step process of making virgin coconut oil, another recorded the method of poultry farming. Each team understood the value of these videos as a catalyst for economic empowerment of women. They recognised that stories could be localised and replicated in different communities according to their own needs, as impacts of climate change continue to affect food resources in island communities.

We will not be moved was produced to highlight the concerns of people in Namata Village in Fiji as they consider the consequences of rising waters on their everyday life and the possibility of relocation to higher grounds. The village chief explains the historical context of the village, which informs viewers about the intrinsic connections between culture and geography. Gender perspectives are included as women describe diminishing fish stocks, their main form of income generation. Mothers describe the dangerous conditions facing their children as they navigate a flooded path to catch the bus to school and express their inability to use the outside kitchen during high tide. Some villagers want to move to higher ground, while others fear that such a move would lead to the loss of important cultural heritage and consequently their identity as a community.

DNA framework as an evaluation tool

The participatory production process enables greater diversity of voices, strengthens networks and empowers participants through the production process. The process can be evaluated using the DNA framework as illustrated in Table 3.4.

The process engenders diversity through cross-fertilisation of ideas between previously disparate groups. Vertical networks initiate dialogue with decision-makers on policy reform, and facilitate knowledge transfer between experts and communities. Horizontal information exchange improves information flow and networking with heterogeneous groups on a range of issues such as adaptation, disaster preparedness and coastal management. It also helps to build consensus and resolve conflict on issues such as mining, water management and environmental damage.

The process of content creation empowers individuals through their newfound agency. Experiential learning and the action-reflection cycle improves participants' capabilities for self-reliance and self-organisation. As participants engage in the production process and begin to understand visual grammar, they also become critically aware of the way

Table 3.4 Evaluating participatory media process using the DNA framework

Is the process facilitating the following outcome:	*Identify what aspects of the production process contributes to the DNA outcome:*
Diversity	
Cross-fertilisation and new connections – intercultural, intergenerational, values and beliefs	Includes diverse networks, knowledge systems and multiple perspectives in content creation.
Inclusion of demographic variables (age, gender, race, identities, education, place disabilities etc.)	Acknowledgement of each person as valuable to the process irrespective of their background. Respect and awareness of the 'other'.
Holistic understanding Consensus building	The process of researching, interviewing, listening, viewing and feedback.
Network	
Heterogeneous and Relational (form and content)	The production process links ideas, skills and knowledge across heterogeneous networks (human and non-human).
Social capital Reciprocity Social cohesion	Use of trusted social networks in the production cycle – planning, collaborating, negotiating, exchanging resources and sharing skills.
Vertical and horizontal networks	Distribution of content upstream. Cooperation and collaboration between endogenous networks.
Agency	
Empowerment Action-reflection-action cycle Self-expression	Taking ownership of storytelling. Awareness of the process, own thoughts and others' needs. Gaining media literacy skills.
Problem-solving Mutual learning Dialogue Participatory action research Critical thinking Self-organisation	Collaboratively discovering new knowledge. Listening and responding to other perspectives. Gathering, analysing and designing messages. Group viewing session and feedback. Dissemination and knowledge sharing.

Source: author revised from Harris (2014)

in which storytelling can be manipulated by the choice of shots and the interview process. They begin to critically reflect on mainstream media's role in influencing public understanding of issues by giving greater voice to political and economic elites. Through a feedback loop of viewing and reviewing the videos, the participants also identify strengths and weaknesses in their own practice and recognise the stock of knowledge and skills available within the group.

Case study 3.3 Community radio-based disaster risk reduction in North Sumatra

By Iwan Awaluddin Yusuf
***Independent project reviewer*[3]**

- *Project:* A community radio network was set up in Aceh to overcome the lack of information and communication after the 2004 tsunami. The Aceh-Nias Radio Network Rehabilitation and Reconstruction of Aceh (ARRNet) community radio programme began on September 2005 and operated until January 2008. The targeted regions of ARRNet were the Aceh and Nias Island (North Sumatra) communities, who were either directly or indirectly affected by the earthquake and tsunami. The main objectives of this programme were to facilitate cross-community conversations and improve communication between the tsunami-affected communities and other stakeholders (government, NGO, funding bodies, volunteers, media, etc.) involved in the rehabilitation and reconstruction process. Community radio is owned by, managed by and dedicated to a community. Community radio provides a place for people to talk, to discuss, to perform art and to express opinion with respect to the common interest. It is also important during disasters for communication exchange.
- *Context:* On 26 December 2004, a tsunami devastated Nangroe Aceh Darussalam (NAD), the westernmost region of Indonesia. As many as 164,000 people died or were reported missing and more than 400,000 people became

homeless. The infrastructure collapsed, including communication networks.

There were 200 humanitarian organisations and 3,000 military personnel from various countries involved in the rehabilitation process. With poor communication infrastructure, coordination and information exchange became very difficult among the organisations and with communities. The aid-targeted groups had difficulty in getting information and updates about what assistance was available to them, how to get the assistance, the value of the aid to which they were entitled, how to participate in these programmes, and where to report when there was a problem. After assessing the situation, the availability of trusted and equitable information was the biggest and most valuable form of aid for the affected communities.

- *Process:* The Yogyakarta based Combine Resource Institution (CRI), with funding support from the Japan Social Development Fund (JSDF) and the World Bank, designed a pilot programme under which five community radio stations in five regions were successfully established. Eventually, 37 community radio stations in nine regions in Aceh and two regions in Nias were introduced under the ARRNet programme. Throughout this process, local communities were consulted about technical and administration needs, capacity building in management and communication networks and the organisation of local people in the management of the stations.

  Each community radio station in Aceh grew according to the local characteristics, culture and customs. The majority of the stations were located in *gampông* (villages) affected by the tsunami. The stations then took on the characteristics of *gampông*-based community radio. Hence the characteristics of each community radio in Aceh varied according to its village. Some community radio stations were also established in non-affected areas.

  After the initial stage, the local citizens took control of the improvement of the community. Generally, the people of

Aceh eagerly welcomed radio as it could accommodate their need for information and entertainment in order to release anxiety and uncertainty after the tsunami. It also empowered them because they could engage actively in the process of rehabilitation and reconstruction through community radio programmes. Community participation was initially very good. People helped to set up the tower and provided entertainment materials and other essential information.

As the years progressed, several challenges emerged as the spirit declined. Local people's active participation, so important at the beginning, was replaced by more passive listening, as people's priorities shifted towards earning a living on a daily basis.

Furthermore, some community radios in Aceh adopted "pop" commercial on-air personalities as young people dominated the management and announcing positions. The managers did not comprehend completely the main principles of community radio. Instead, success was equated to the number of advertisements on the radio. Managers also struggled to understand and comply with institutional issues such as licencing and technical problems, which took up much of their time. Many of the stations have closed down because of these problems and the lack of external technical support.

- *Outcome:* The ARRNet programme is a good example of the application of Community Radio-Based Disaster Risk Reduction or CRB-DRR concept. CRB-DRR emphasises the role of community radio during three stages – before disaster for disaster mitigation and preparation; during disaster for early warning systems and evacuation information; and after disaster for problem mapping and aid distribution. The main aim is to facilitate the exchange of information in both directions between communities affected by a disaster and other stakeholders (government, NGO, funding, volunteers, media, etc.).

  ARRNet operated in the post-disaster stage. The programmes centred on post-tsunami rehabilitation and reconstruction activities in Aceh by meeting community needs for information. Collaboration between community radio, aid agencies, and the community acted as a public communication tool to exchange easily understood information quickly and accurately. The ARRNet programme is an example of community radio's role in disaster information management.

Message dissemination

User-generated content (UGC) is easily distributed to international audiences or to networks within minutes of completion through easy file-sharing options via the Internet and without the need for approval of a gatekeeper. However, it is crucial to ensure that members of the community who participate in the message-making process give their consent for distribution outside of the community. Participants must be consulted and made aware of the potential repercussions of having their work disseminated on social media. Factors that influence this decision include relevance and sensitivity of the topic, safety of those appearing on camera, especially in conflict situations, cultural taboos and intellectual property concerns.

The decentralised nature of participatory media activities means that many of these projects have a short life span and are disseminated in ways that may not reach the target audience.

Sharing content online does not necessarily mean that it will get to the target audience. For example, if one of the objectives of the exercise is to reach decision-makers, strategies must be developed to share the message with relevant politicians and organisations in face-to-face meetings and conferences. Exchanges can be facilitated within the community using community screening, emails or direct messaging apps on mobile phones.

Research and ethical issues

There are several ethical issues to consider when engaging with participatory media activities using video and photography. These can be looked at in relation to research and consent. The standard way to gain consent is to use protocols developed by university ethics committees as part of the ethics approval process. This includes a standardised written consent form used to gain individual consent from research participants. It provides information about the project and contact details of the facilitator, and gives people the option to withdraw at any stage of their participation. The form is signed by all research participants who are interviewed. If an audio or video recording is made of the interviewees, it should be clearly explained to the participants how the organisation or researcher plans to use the material. Participants give their consent for its distribution. Participants must also be made aware of the risks arising from their attendance in workshops and from the use of the images, especially in conflict situations (Harris & Bau, 2014).

Increasingly, practitioners are considering other, more culturally appropriate forms of consent which can be recorded in audio-visual form either at the start or the end of the interview recording. This allows participants who don't have reading and writing skills to be able to clearly articulate their consent and any restrictions on the use of the material. In some communities, communal consent may be gained through a protocol which receives the chief's or elder's approval to film within traditional sites, villages and settlements.

Verification of UGC is necessary in order to guard against the spread of rumours and misinformation in emergency situations. Verification Handbook[4] (Silverman, 2014) offers useful guidelines for verifying UGC and open source information in an era of fake news and crisis journalism. It includes relevant case studies, a verification process, checklist and tools to support a professional or citizen journalist.

Conclusion

Participatory environmental communication provides a viable discursive space in which to engage communities about a range of environmental issues. Participatory media improves collective action and knowledge exchange between diverse stakeholders across different networks; intercultural, intergenerational, interdisciplinary, interagency and cross-sectoral. Collaboration encourages transactional communication and reciprocity through the production process. Enabling agency of a community means providing opportunities for members to engage in collective dialogue to find solutions which improve their community and consequently improve their own lives. It asks participants to imagine a different reality. To achieve this, people need access to reliable information and the ability to act upon it collectively.

Transactional communication places great importance on how meaning is constructed in one's mind and upon reflective feedback and reflective listening as demonstrated by the Wan Smolbag Turtle theatre project. Participants' active engagement in communicative processes increases knowledge exchange and improves adaptive capacities as seen in Andra island photo essays. Community media provides a dynamic model of engagement which reflects the local context as discussed in the ARRNet community radio project. These case studies demonstrate that inviting people to participate in media production improves resilience of the whole community. It improves people's ability to identify the networks and resources available to them and helps them to make that critical link between vulnerability and the necessity to adapt and change.

By studying the process, we begin to understand what elements of people's involvement in media production are empowering, the way they go about determining their message production and how they reflect their worlds through mediated communication. Through these skills, vulnerable communities are better able to communicate and represent their interest in national policy development and international fora. Most importantly, participatory media enable subaltern narratives to become alternative nodes of knowledge creation.

Notes

1 Summarised with permission of authors: Inamara, A. & Thomas, V. 2017. Pacific climate change adaptation: The use of participatory media to promote indigenous knowledge. *Pacific Journalism Review*, 23, 113–132. The study was part of a project implemented by the Wildlife Conservation Society and the Centre for Social and Creative Media, University of Goroka, Papua New Guinea
2 www.youtube.com/channel/UCSegmt_jV8Yjw9-U94NZziw?view_as=subscriber)
3 Summarised from: Yusuf, I. A. 2017. Aceh community radios in disaster information management: Risk reduction, resilience and preparedness. The 3rd World Conference on Media and Mass Communication 2017 (MEDCOM 2017), 20–22 April, 2017, Kuala Lumpur, Malaysia.
4 http://verificationhandbook.com/book/

References

Adler, R. B. & Rodman, G. R. 2003. *Understanding Human Communication*. New York: Oxford University Press.

Arnstein, S. R. 1969. A ladder of citizen participation. *Journal of the American Institute of Planners*, 35(4): 216–224. doi:10.1080/01944366908977225

Carpentier, N. 2016. Beyond the ladder of participation: An analytical toolkit for the critical analysis of participatory media processes. *Javnost-the Public*, 23(1): 70–88. doi:10.1080/13183222.2016.1149760

Chin, S. Y. 2000. Development communication in Asia. *In:* Servaes, J. (ed.) *Walking on the Other Side of the Information Highway: Communication, Culture, and Development in the 21st Century*. Penag: Southbound.

Díaz-Bordenave, J. (1989). *Participative Communication as a Part of Building the Participative Society*. Paper presented at the A Key Concept in Communication for Change and Development, University of Pune, India.

Faris, S., Lipscombe, S., Whitehead, S. & Wilson, D. 2014. *From the Ground Up: Changing the Conversation about Climate Change* [Online]. London, UK: BBC Media Action. Retrieved from http://downloads.bbc.co.uk/mediaaction/policybriefing/bbc_media_action_from_the_ground_up_climate_change.pdf

Freire, P. 1970. *Pedagogy of the Oppressed*. New York: Herder and Herder.

Galbraith, M. 1991. The adult learning transactional process. *In:* Galbraith, M. (ed.) *Facilitating Adult Learning: A Transaction Process*. Malabar: Krieger Publishing Co.

Gaventa, J. & Barrett, G. 2012. Mapping the outcomes of citizen engagement. *World Development*, 40(12): 2399–2410. doi:10.1016/j.worlddev.2012.05.014

Haraway, D. 1988. Situated knowledges: The science question in feminism and the privilege of partial perspective. *Feminist Studies*, 14(3): 575–599.

Harris, U. S. 2008. *Transforming Images: Participatory Video and Social Change in Fiji*. Doctor of Philosophy. Sydney: Macquarie University.

Harris, U. S. 2014. Communicating climate change in the Pacific using a bottom-up approach. *Pacific Journalism Review*, 20(2): 77–96. doi:10.24135/pjr.v20i2.167

Harris, U. S. & Bau, V. 2014. Harnessing participatory media technology for community development: Choices and implications for practitioners. *Development Bulletin*, 76: 31–34, ANU.

Haynes, K. & Tanner, T. M. 2015. Empowering young people and strengthening resilience: Youth-centred participatory video as a tool for climate change adaptation and disaster risk reduction. *Children's Geographies*, 13(3): 357–371. doi:10.1080/14733285.2013.848599

Kalas, P. P. & Finlay, A. 2009. *Planting the Knowledge Seed-Adapting to Climate Change Using ICTs: Concepts, Current Knowledge and Innovative Examples* [Online]. Melville, South Africa: Building Communication Opportunities (BCO) Alliance. Retrieved from www.apc.org/sites/default/files/BCO_ClimateChange.pdf

Kothari, U. 2001. Power, knowledge and social control in participatory development. *In:* Cooke, B. & Kothari, U. (eds.) *Participation: The New Tyranny*. New York: Zed Books.

Lennie, J. & Tacchi, J. 2013. *Evaluating Communication for Development: A Framework for Social Change*. Oxon: Routledge.

Liao, H-A. 2006. Toward an epistemology of participatory communication: A feminist perspective. *The Howard Journal of Communications*, 17(2): 101–118. doi:10.1080/10646170600656854

Lunch, N. & Lunch, C. 2006. *Insights into Participatory Video: A Handbook for the Field*. Oxford: InsightShare.

Mansuri, G. & Rao, V. 2013. *Localizing Development: Does Participation Work?* Washington, DC: The World Bank.

McNamara, K. E. 2013. Raising awareness about climate change in Pacific communities. *Environmental Education Research*, 19(6): 864–871. doi:10.1080/13504622.2013.769046
Melkote, S. R. & Steeves, H. L. (2001). *Communication for Development in the Third World: Theory and Practice for Empowerment*. London: Sage.
Norris, P. 2002. Social capital and the news media. *The Harvard International Journal of Press/Politics*, 7(1): 3–8. doi:10.1177/1081180x0200700101
Phillips, L., Carvalho, A. & Doyle, J. 2012. *Citizen Voices: Performing Public Participation in Science and Environment Communication*. Bristol, UK: Intellect Bristol.
Phillips, L. J. 2011. *The Promise of Dialogue: The Dialogic Turn in the Production and Communication of Knowledge*. Amsterdam, The Netherlands: John Benjamins Publishing Company.
Piccolella, A. 2013. Participatory mapping for adaptation to climate change: The case of Boe Boe, Solomon Islands. *Knowledge Management for Development Journal*, 9(1): 24–36.
Pileggi, M. & Morgan, E. 2017. Storytelling as action: Exploring transformation. *In:* Milstein, T., Pileggi, M. & Morgan, E. L. (eds.) *Environmental Communication Pedagogy and Practice*. London: Routledge.
Plush, T. 2009. *Video and Voice: How Participatory Video Can Support Marginalized Groups in Their Efforts to Adapt to a Changing Climate*. Sussex, UK: Institute of Development Studies.
Saifoloi, M., Papoutsaki, E., Williams, M., Harris, U. S. & Naqvi, M. 2016. Participatory video and the Pacifica Mamas: A pilot project. *Unitec ePress Research Report Series*, 2: 1–20.
Shaw, J. & Robertson, C. 1997. *Participatory Video: A Practical Approach to Using Video Creatively in Group Development Work*. London: Routledge.
Servaes, J. 1999. *Communication of Development: One World, Multiple Cultures*. Cresskill, NJ: Hampton Press.
Silverman, C. 2014. *Verification Handbook: A Definitive Guide to Verifying Digital Content for Emergency Coverage*. Maastricht, The Netherlands: European Journalism Centre.
Spurgeon, C. 2015. Participatory media and "co-creative" storytelling. *Media International Australia*, 154: 132–138. doi:10.1177/1329878X1515400116
Tufte, T. & Mefalopulos, P. 2009. *Participatory Communication: A Practical Guide* [Online]. Washington, DC: The World Bank. Retrieved from http://siteresources.worldbank.org/EXTDEVCOMMENG/Resources/Participatory communication.pdf
Uslaner, E. M. & Conley, R. S. 2003. Civic engagement and particularized trust: The ties that bind people to their ethnic communities. *American Politics Research*, 31(4): 331–360. doi:10.1177/1532673X03031004001
White, S. A., Nair, K. S. & Ascroft, J. 1994. *Participatory Communication: Working for Change and Development*. New Delhi: Sage Publications.

Part II

Bringing Pacific Island perspectives

4 Dialogic encounters

Listening to people's stories

Reflection

More than 97% of scientists who study climate science, and other categories of scientists, agree that human-induced global warming is happening (Cook et al., 2013; Anderegg et al., 2010; Oreskes, 2018). Their research points us towards one thing – our earth is warming and will continue to warm if we don't reduce our reliance on fossil fuels. It may be difficult for lay people to understand all the graphs and data on climate change produced by scientists, and how these changes in the climate affect all species on earth. How does a warming planet affect people at a personal level? How does it translate into people's everyday lives? To gain an insight into this, I travelled to Pacific island countries that have the smallest carbon footprint in the world yet stand to endure the greatest damage from the impacts of climate change.

> People who are protected by their wealth have little idea of the heartbreak of the poor and most vulnerable when they lose their homes and their livelihoods in climate-related disasters.
>
> Frank Bainimarama, Prime Minister of Fiji and COP23 President[1]

Introduction

In this chapter I share stories of the people I met in Fiji, Vanuatu and Kiribati, communities who are experiencing changes to their environment, and others working alongside these communities in adaptation and disaster response. In Kiribati, I learned about a family's concerns of an insecure future caused by coastal erosion which threatens their ancestral land and family home. In Fiji, I listened to the story of a subsistence farmer whose livelihood has deteriorated with the onset of extreme weather events. In Vanuatu, I learned about the importance of communication and collaboration in the face of natural disasters. Above all, I discovered the indomitable spirit and perseverance of both citizens and governments in Pacific island countries (Figure 4.1) striving to build resilient societies from the ground up in the face of a changing reality.

Talanoa

Stories are agents of change. Learning from the stories of affected communities builds empathy and invites us into their world. Many of the stories shared in this chapter were received during informal talanoa sessions.[2] Talanoa is a Pacific communicative practice in which people come together to share stories and to discuss important issues.

Pacific island academic and now Tongan politician, Sitiveni Halapua, has been instrumental in developing the theory and practice of the talanoa approach. He has taken it from the village setting to national and international forums. Halapua explains: "Talanoa does not have a preconceived agenda. It is very open; you can tell your story. Prior to the advent of Western civilization and the coming of the missionaries, talanoa was how our history was created" (SGI Quarterly, 2007: 9). Robinson and Robinson (2005: 18), who have studied this process, explain it further:

> By listening to and, hopefully, understanding those who have differing ideas, views or opinions the intention is that talanoa will lead towards a common point of consensus. People need not necessarily agree entirely with each other; what is important within talanoa is that they have listened and understood the position of the others involved.

It was this idea of bringing all parties together in respectful dialogue that was adopted at the United Nations climate talks in 2017 during Fiji's presidency of COP23, at the suggestion of Fiji's Prime Minister and COP23 President, Frank Bainimara.

My talanoa sessions with communities were organised in advance by a local intermediary. The conversations themselves were open-ended and led by the speaker. Questions were asked in response to the story being told by each speaker, mainly to extend my own understanding of the topic. The talanoa sessions were only recorded if the speaker gave his or her consent. Where translations were necessary the local intermediary took up the role of both the interlocutor and translator. The

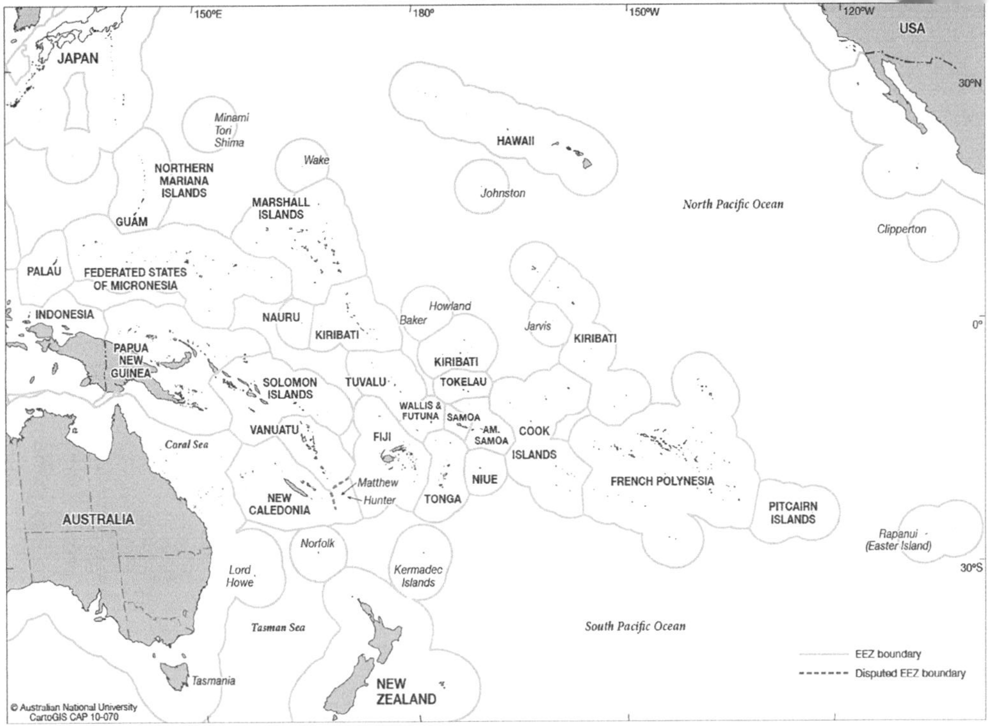

Figure 4.1 Map of Pacific Islands region showing the Exclusive Economic Zone (EEZ) boundary for each country

Source: Map reproduced with the permission of CartoGIS Services, ANU College of Asia and the Pacific, The Australian National University.

talanoa took place on location to include visual references and enhance my understanding of the context in which people lived. In some of the sessions, other adults and children joined the conversation by sitting alongside the speaker, listening and occasionally contributing.[3]

Kiribati

Kiribati is located near the equator in the central Pacific Ocean between approximately 5°N—12°S and 168°E—152°W. The country consists of 33 low-lying atolls.

Population: 110,136
Median Age: 25.5 years
Life Expectancy: Women: 66 years, Men: 58 years
Gender: Male: 49%; Female: 51%
Literacy: 92 %
Language: English, I-Kiribati (or Gilbertese)
Urbanisation: 48.7 % of total population
GDP per capita: $1573 (AUD)
Access to mobile network: 48.5 per 100 inhabitants with mobile cellular subscription. 60% of the population has 3G coverage.
The Exclusive Economic Zone: 3.6 million km^2

Vulnerabilities:

- Impact of sea level rise on agriculture, coastal erosion and fresh water supply
- Extreme drought, tsunami threat
- Frequent storms, coastal flooding, longer droughts
- Sea level rise of 1–4mm per year since 1993.

Source:

1. Kiribati National Statistics 2015 Census Report
2. ITU, Country profile Kiribati 2017
3. Climate change information: Australian Bureau of Meteorology and CSIRO, 2011. Climate Change in the Pacific: Scientific Assessment and New Research. Volume 1: Regional Overview. Volume 2: Country Reports. Accessed from www.pacificclimatechangescience.org/publications/reports

Kiribati

> It is our moral obligation, individually and collectively, to continue to remind the global community that unless resources are mobilized and targeted to where it is needed the most, the fate of small island states captures the essence of what global warming might mean at its most severe. Their security is threatened. Their survival, as a nation, as a people and as a cultural identity is on the line.[4]
>
> (Anote Tong, former President of Kiribati (2016))

The small six-seater plane lifts off the runway that spans the widest part of the island in South Tarawa, Kiribati. As the plane travels skyward I look down at the serpentine land below, barely above the water, with a lagoon on one side and the ocean on the other. The plane is hardly in the air before it begins its descent towards Abaiang, another low-lying coral atoll. An hour-long bumpy lorry ride and we are in Tebontebike, a village on the southern end of the island.

This story belongs to a 65-year-old woman named Maria Tekaie. Maria leads us towards the sea to a place where her family once lived. The rocky outcrops seen several metres out in the sea were once part

Figure 4.2 Maria Tekaie at Tebontebike, Abaiang Island

Source: Usha S. Harris

of the road which encircled the point where she brought up her children (See Figure 4.2). The relentless sound of waves carving out the shore is ever present in the background as Maria tells me her story.

> ***The sea is eating my land***
>
> *We used to live in another village. The British people put everyone in one village. But that village . . . the land belongs to other people. Gradually the owners of the land claimed their right so we have to return to our place, which is here. Many times now we've been moving inland. We love to live near the sea because we get the breeze and it's easy to go fishing.*
>
> *It makes us scared living near the sea because the sea keeps taking our land. Lots of time we've moved. The kitchen, I said to the children, don't make it permanent. It has to be easy to carry, don't build it here now because we will have to move further in.*
>
> *Do you know why this is happening?*
>
> *I don't know why it is happening because before, during my youth, nothing really was happening.*
>
> *Who can you approach for help?*
>
> *We asked the council, why they don't come and fix this because we're worried we'll have no land. They put up a sea wall, but the sea is stronger than their sea wall. And now we can't plant our food in the pit because the sea comes right in.*
>
> *Do you know what is climate change?*
>
> *I heard it on the radio, But I wonder what is the cause of it? We heard about it but we don't know what it is. We know we are losing our land and already we have lost a lot of our land. We have to move because this place is not going to grow our food. We have some taro pit in the other village and that is where we get our food from.*
>
> *We love this village, our village. We already lost 100 metre of land. If we lose our land maybe we'll have to move to some other part of the island. But we just love here, we just can't . . . We're very sad because we know now that the sea will never stop eating our land. I'm Christian and I will place hope in God's hand that*

he will look after it and whatever happens we will accept, but now as we experience this it makes us scared of what will happen.

Are there any traditional stories which tell of such an event in the past?

I can't remember any incident that the land is eaten away. We just enjoyed our land because everything was normal but now, just in 10 years, it's become very strong. There used to be a bay there and a road but now it's all gone.

(M Tekaie, 2017, personal communication, 29 July)
[Translation: Claire Anterea]

Freedom to choose

By listening to Maria's story we begin to understand how history and politics of the past continue to ripple into the present and inform her sense of security. Paulo Freire's emancipatory pedagogy, where the process of action and reflection (praxis) leads the oppressed towards transformation, is especially relevant here:

> Human existence cannot be silent, nor can it be nourished by false words, but only by true words, with which men and women transform the world. To exist, humanly, is to *name* the world, to change it. Once named, the world in its turn reappears to the namers as a problem and requires of them a new *naming*. Human beings are not built in silence, but in word, in work, in action-reflection.
> (Freire, 1970: 88)

In naming their world, ordinary people like Maria reflect the true nature of their situation through their own lived experiences, instead of the hegemonic political debates framed in global media. Maria and her family do not want to be displaced from their land again. She has experienced displacement once before during British colonial rule when many families on Abaiang Island were relocated. Living on their own land gives them a sense of agency, which was taken away from Maria's family when she was young. Her love for the sea is a significant aspect of her own cultural identity.

The ecological identity of the I-Kiribati (people of Kiribati) is closely linked to the sea and embedded in their stories, songs and dance. People hold great respect and love for the ocean. Claire Anterea, from

Kiribati Climate Action Network (KiriCAN),[5] an environmental NGO which is on the frontline of the fight to save coastal ecosystems, observes:

> The ocean is our mother. We get most of our food from the ocean and we love to swim in the ocean. We are always surrounded by the sea. It's always part of us. . . . But lately, now, it's become a threat to us, the mother that we used to know and love. It looks like it's an angry mother. Every time when there is a high tide, and every time there's a storm, we know that seawater will go into people's homes. Seawater will go into our wells . . . It's become something scary.
>
> (C Anterea, 2017, personal communication, 02 August)

We can evaluate Maria's situation using the capabilities approach, which prioritises the expansion of individuals' capabilities and freedoms. Amartya Sen (2013: 6) argues that understanding sustainability within a 'needs' approach as framed in the Brundtland Commission Report (WCED, 1987) is incomplete. Rather, he proposes, "A fuller concept of sustainability has to aim at sustaining human freedoms, rather than only at our ability to fulfil our felt needs" (Sen, 2013: 6). For him, agency of individuals is vital in leading a sustainable life. In Sen's (1999) capabilities approach, functionings are 'beings and doings', that is, various states of a human being and activities that a person can undertake. Maria's functionings can be identified as follows (in Table 4.1).

What would make Maria's life better? She needs greater choices, which would allow her family the freedom to live on their land if they choose to stay instead of facing the prospect of displacement again. Maria emphasises that she loves the sea, but she does not

Table 4.1 Maria's functionings

Beings (what Maria can be)	*Doings (what Maria can do)*
Insecure	Plan for a secure future
Homeless	Live in a permanent home
Malnourished	Have food security
Afraid	Live by the sea that is eroding her land
Happy	Live by the sea which provides joy and sustenance to her family
Empowered	Have a voice and choice

Source: author informed by Sen (1999, 2013)

understand why the sea is claiming her land. Coastal erosion is a complex issue caused by a multitude of factors including both natural processes (waves, water currents, stronger wave action due to stronger cyclones) and human activity (illegal sand mining, over development, use of ineffective defences such as sea walls) (Paeniu et al., 2015).

An expansion of choices would include technical and financial assistance that could provide coastal communities with a viable solution, allowing them to continue to live on their traditional lands and by the sea, which she loves. Pacific island governments are developing relocation guidelines which include respecting voluntary immobility when a community decides not to move (Farbotko, 2018).

The I-Kiribati decision-makers emphasise that adaptation is about giving people choices. This means assisting those who want to migrate to gain new skills so they can migrate with dignity, and helping those who want to stay in Kiribati to live at home enjoying the rights to clean water, food security, and safe housing. This requires the affected communities to gain the skills to grow food and build strong structures and safe homes, which would withstand increasingly severe weather events. But to do this Kiribati needs the support of the international community. It needs specialists who can share their skills and knowledge with local people to build a strong infrastructure, maintain sustainable agriculture, implement adaptation methods supported by science and local knowledge, and have access to better communication. Kiribati must have access to climate finance to build the adaptive capacity for current and future generations.

Climate justice

Climate justice refers to moral and/or ethical consideration of addressing climate change, rather than approaching it just as an environmental and social problem. Climate change impacts are unequivocal, and must be addressed with immediate effect by developed and developing nations. Page (2006: 3) stresses the importance of applying ethics to address climate change impacts. A form of this is "distributive justice[6] or the study of how benefits and burdens should be distributed across space and time". Such an approach would ensure a fairer deal for low-lying developing and small-island states who carry a disproportionately higher social and economic burden from the impacts of climate change, and achieve intergenerational equity for future generations. Thus, additional support is essential for those vulnerable nations who are less responsible for global GHGs emission.

When 195 countries reached the Paris Agreement, the Green Climate Fund (GCF) was given an important role in serving the agreement and supporting the goal of limiting global warming to well below 2 degrees Celsius (Green Climate Fund, 2017; Hewitt et al., 2012). Least Developed Countries, Small Island Developing States and African states are most vulnerable to climate change. They want access to the GCF because it supports their efforts to limit the effects of climate change and to build greater resilience among their populations. Access to the fund by vulnerable nations would help to minimise climate change impacts and reduce global GHGs emission.

However financial viability of this fund is still uncertain due to United States' refusal to meet their outstanding contribution (Schalatek, 2017). Lack of political will in climate initiatives and climate justice issues prompts many groups and environmental NGOs to engage in the climate justice movement in support of vulnerable communities (Pettit, 2004; Abate, 2010; Klein, 2014).

The Catholic NGO Caritas that works closely with vulnerable communities in the Pacific has found that climate finance is inadequate in all three categories: quantity needed on an annual basis, overall quality of finance in the area of adaptation, and support for the most vulnerable people who face uncertainty "over the security of their local food and water sources, future weather, and the long-term viability of their homelands" (Caritas Oceania, 2017: 11).

Industrialised countries that are responsible for the largest amount of greenhouse gas emissions have also benefited the most economically for more than a century. They are being asked to take greater responsibility in assisting people like Maria and her family to adapt. Linda Uan, a veteran I-Kiribati journalist and media producer who has seen the impacts of a warming world first-hand while working with remote communities in Kiribati, understands the importance of climate finance:

> In Kiribati culture when you do a wrong, when you make a mistake, when you cause a damage, you repair it. . . . Our people are saying we didn't break it but we are having to live with it. You are telling me I can't laugh anymore, I can't joke anymore, I can't go and fish and do what I like because I have to get up and fix the damage that's been caused by the white men? . . . It is about our very survival. . . . We are not saying give us money so that we can buy more air conditioning. We want to live here.
>
> (L Uan, 2017, personal communication, 2 August)

Hope and renewal

Unlike the cataclysmic portrayal of the country in international media, the people in Kiribati are fighting a battle for survival of their ancestral land. More than 50% of the population is under 24 years of age. I-Kiribati are great travellers despite the remote location of the islands and infrequent airline service. Many young people leave to study in Fiji and gain their first work experience overseas. As experienced seafarers, many I-Kiribati men work as crew for international shipping companies. However, some young people are returning after gaining their education overseas to seek solutions which will allow them to continue living in their homeland. They are focused on finding sustainable ways of turning to productive use the resources of the ocean and the thin sliver of the atoll on which they live. It is the voices of young I-Kiribati which bring hope to an otherwise bleak narrative because they are the agents of change. For this reason, it is also important to share the story of hope and renewal in Kiribati.

Keboua John Collins is one of these inspiring young I-Kiribati living on the island of Abaiang. Although his parents migrated overseas, Keboua returned to Abaiang Island to fulfil his father's dream of building beach bungalows: "That's my dream. I want to go back to my home island. I want to work closely with the community. I want to help our people", he tells me. In two years he has transformed his land from scrub into an eco-lodge. Terau Beach Bungalows was built using traditional design principles and using local materials such as coconut trunks and pandanus leaves which are found in abundance on the island.

An agricultural specialist, Keboua received part of his training in South Korea, where he learned about the New Village Movement known as Saemul Undong, a community-led development plan introduced by President Park Chung-Hee to rebuild the society after the Korean war (see The Movement for Community-Led Development, 2017). Influenced by this Korean cultural practice of self-sustainable community building, Keboua sought help from village elders to teach unemployed youth about sustainable building practices which were being lost. This has aided intergenerational knowledge transfer and capacity building for unemployed youth.

Keboua also teaches communities about making compost using local organic material to build up the soil content. This allows villagers to grow vegetables and reduce reliance on imported foods. One of the challenges is access to good quality seeds. He describes his way of working with the Abaiang community as the "less approach – how to work with the available resources to engage or develop the community". His relationship with the community is based on trust and fair exchange known as *Bubuti*, a form of reciprocity.

We find all the attributes of the *DNA* framework in this story. Keboua uses scientific knowledge that he has acquired as an agricultural specialist to teach others on the island about composting. In an intercultural exchange, he has combined the cultural knowledge of his own people with the South Korean community-building practice of Saemul Undong to find a sustainable development model for his island community. He uses his kinship network to enact intergenerational knowledge exchange, at the same time catalysing the agency of others in the community through group dialogue.

Traditional knowledge in Kiribati is highly protected within each family and is passed orally to the next generation only if a young person is considered trustworthy. It is prohibited (*tabu*) to share one's knowledge with other families or outsiders. Many forms of oral traditional knowledge have been lost because of this practice, coupled with the fact that the older generation is dying prematurely because of non-communicable diseases. After climate change, this is the most serious issue facing Kiribati. Non-communicable diseases account for around 70–75% of all deaths in the Pacific Islands (Anderson, 2013).

Fiji

Fiji is located in the western South Pacific Ocean between 177°E—178°W and 16°S—20°S.The country consists of 322 islands.

Population: 885 000
Median Age: 27.5 years
Life Expectancy: Women: 73 years, Men: 67 years
Gender: Male: 50.7%; Female: 49.3%
Literacy: 93.7% (Male: 95.5%; Female: 91.9%)
Language: English, iTaukei (Fijian), Fiji Hindi
Urbanization: 55.9% of total population
GDP Per Capita: $7685.9 (FJD)
Access to mobile network: 116.3 per 100 inhabitants with mobile cellular subscription. 96% of the population has 3G coverage.
The Exclusive Economic Zone: 1.3 million km^2.

Vulnerability:

- Average of 16 cyclones per decade
- Droughts associated with El Niño events
- River flooding every wet season
- Storm surges and waves associated with significant tropical disturbances
- Sea level rise of 6 mm per year since 1993

Sources:

1. Fiji Bureau of Statistics 2017 Census
2. ITU, Country profile Fiji 2017
3. Climate change information: Australian Bureau of Meteorology and CSIRO, 2011. Climate Change in the Pacific: Scientific Assessment and New Research. Volume 1: Regional Overview. Volume 2: Country Reports. Accessed from https://www.pacificclimatechangescience.org/publications/reports/

Fiji

I arrive in Fiji six months after Tropical Cyclone Winston, the strongest cyclone recorded in the southern hemisphere, ravaged the islands. TC Winston killed 44 people, left tens of thousands of people homeless, destroyed schools, health facilities and communication infrastructure and shrank Fiji's economy by a third. The small rural town of Rakiraki, on the north of the main island of Viti Levu, was caught in the eye of the storm. My nephew lives in the neighbouring town of Tavua. His children share frightening memories of that night when their house swayed and their roof 'heaved' as 'fearsome' winds caused ruin and destruction around coastal villages and towns. We visit inland farming communities who have cobbled together a makeshift kitchen here or a bedroom there while waiting for government assistance to arrive. With their farms destroyed there is no money to rebuild the houses. We sit under the shade of a mango tree as farmers Ajay Kumar and Ranika Devi share their stories, occasionally pointing out the devastation to their farm.

When the earth cracks open

Ajay: *I was born in Viti Vanua. My grandfather began farming this land. When I was small I remember we grew peanuts, corn and lentils. When I was 11 years old my father began sugar cane farming. I have continued farming sugar cane. We don't have as much rain as we used to. We mainly rely on sugar cane for our income. To grow market vegetables like long bean, pumpkin, peanuts requires a steady supply of water and three times as much labour.*

Ranika: *Before we used to sell these vegetables to gain extra income but now we have to buy vegetables from the market for our own needs.*

Ajay: *The drought was extremely bad last year. All our cane dried up. It wasn't only our crops. We did not even have enough drinking water. When the earth cracks open then nothing can grow. After the drought we were badly affected by TC Winston. We lost 90% of our crops. Now six months later we only have 30% yield. If the weather does not support us then it is very difficult for us to make ends meet. In the past, I used to cut 400 tons of cane, now I harvest only 100 tons. The situation is dire. The cost of fertiliser is also very high. My family's future will become more difficult. We have to work extraordinarily hard to survive.*

Ranika: *When the rainfall was good each watermelon weighed around 8 to 10 kilograms. But now the watermelons which grow in the same field, on the same spot are no heavier than 2 to 3 kilograms. The taste and colour have changed as well. The mango trees were also badly damaged in the cyclone. It will be 5 to 6 years before we can enjoy the same amount of mango harvest as we did before the cyclone. We were able to supplement our income by selling mangoes. Life is very hard now. We have gone backwards by more than 10 years in terms of our ability to farm and earn a good income. I don't know what will happen to us in the future.*

(A Kumar & R Devi, 2017, personal communication, 14 August)

Facing climate realities

I began to understand how extreme weather events trap families in a spiralling poverty cycle. By listening to the stories of farmers, I realise that the graphs and charts illustrating changed rainfall patterns or prolonged droughts do not tell the whole story. I think back to the stories

I heard as a child from my mother about the farming families caught in floods. The generations have changed but the stories remain the same. The increasing intensity and frequency of these floods, cyclones, droughts, heatwaves and super storms have enormous impact on communities that live a subsistence lifestyle.

Before TC Winston, an El Niño-induced drought affected up to 30,000 people in Fiji. The Fiji government provided emergency water supplies to many rural and outer island communities as the water supply dried up. El Niño-related severe drought conditions were experienced by many other Pacific island countries, including Tonga, Solomon Islands, Marshall Islands, Papua New Guinea and Kiribati. Some of these countries declared a state of emergency as they ran out of fresh water sources leading to crop failures and food shortages (ReliefWeb, 2018). Robert Glasser, head of the UN Office for Disaster Risk, writes "About 90 percent of recorded major disasters are now weather and climate related. Extreme weather has played a role in the first rise in 10 years in the numbers of people who go hungry every day" (UNISDR, 2017).

Vanuatu

Vanuatu lies in the South Pacific Ocean between 13°S—21°S and 166°E—171°E. The archipelago consists of 83 islands.

Population: 272, 459
Median Age: 20 years
Life Expectancy: Women: 73 years, Men: 70 years
Gender: Female: 49%; Male: 51%
Literacy: K-12 school enrolment was 80,800 students in 2015
Language: Bislama, French, English
Urbanisation: 25% of total population
GDP Per Capita: 4,455 Vt. (Based on average total weekly expenditure per capita adult equivalent)
Access to mobile network: 81.1 per 100 inhabitants with mobile cellular subscription. 80% of the population has 3G coverage.
The Exclusive Economic Zone: 710 000 km^2

Vulnerability

- An average of 23 cyclones per decade
- Drought and coastal and river flooding related to La Niña and El Niño events

- Effect of acidification on the health of reef ecosystems compounded by other stressors including coral bleaching, storm damage and fishing pressure
- Earthquakes, landslides, tsunamis and volcanic eruptions

Sources:

1. Vanuatu National StatisticsOffice
2. ITU, Country profile Vanuatu 2017
3. Climate change information: Australian Bureau of Meteorology and CSIRO, 2011. Climate Change in the Pacific: Scientific Assessment and New Research. Volume 1: Regional Overview. Volume 2: Country Reports. Accessed from www.pacificclimatechangescience.org/publications/reports

Vanuatu

Vanuatu is one of the most disaster-prone countries in the world affected by natural disasters such as cyclone, tsunami, earthquake and flooding. Vanuatu's vulnerability is further worsened because of its location in the Oceania region, which has the highest disaster risk of all regions in the world (Bündnis Entwicklung Hilft, 2017). So, it is no surprise that a society-wide dialogue is taking place about building resilience in this archipelago of 83 islands.

The World Risk Index (Bündnis Entwicklung Hilft, 2017: 8) looks at the exposure of a country based on the following four factors:

1. Exposure to natural hazards such as earthquakes, hurricanes, flooding, drought and sea-level rise
2. Vulnerability of infrastructure, nutrition, living conditions and economic circumstances
3. Coping capacities of governance, preparedness and early warning measures, access to healthcare, social and material security
4. Adapting capacities with respect to impending natural events, climate change and other challenges.

A network of organisations is working to reduce the vulnerability of ni-Vanuatu society after the devastation caused by Cyclone Pam in 2015. The most powerful cyclone in Vanuatu's recorded history impacted 80%of the ni-Vanuatu population, made a quarter of the population homeless and destroyed most of the food crops that rural communities depend on for their livelihood (SPC, 2017). The cyclone hit the same areas of Vanuatu that had been severely affected by droughts in the previous year (Oxfam, 2015).

Fred Samuel[7] was the Government Chief Information Officer when Cyclone Pam struck Port Vila, the capital of Vanuatu, on Efate Island. He explains how, in the midst of the worst disaster, an alliance formed between government, the private sector and civil society organisations to establish emergency communication links so that relief work could resume.

Communication is vital

> Before Cyclone Pam arrived we decided to use SMS facility [to send cyclone warning alerts]. Almost 94%of the population had access to mobile coverage. So it was the most powerful medium. The telecom operators agreed to freely disseminate the information. It was highlighted as one of the top reasons why we had a low death rate. The information was in Bislama [national language of Vanuatu] and referred to the cyclone tracking map which is in every household in Vanuatu. The tracking map shows the islands on grid lines and in our SMS we sent the coordinates, e.g. cyclone is heading towards H10.
>
> As the cyclone approached we discussed how we should collaborate. It was a clear demonstration of a very good, strong partnership between the private sector and the government. After the cyclone everything was destroyed. Central and southern parts of Vanuatu were mainly affected. Communication with the islands to the north, which were unaffected by the cyclone, was also lost because all the main telecom switches are based in Port Vila.
>
> In order for government to respond effectively we need to communicate. In order for NGO's to serve the people we need communication. I think everyone recognised that communication is vital. We received a lot of help from the UN, Microsoft, Cisco, British Telecom, which provided us with emergency technical support. Within 7 days after the cyclone we were able to have 30–40%of communication back online. What we learned from TC Pam is that we need to have an effective communication plan for disaster. I think HF radio is #1 technology when it comes to disaster despite having social media, despite having Internet, Wi-Fi and all the latest mobile technology. Medium wave AM radio broadcast and short wave are very powerful mediums and they are much more resilient technology than other forms of technology.
>
> And don't forget word of mouth. It has been here with communities for thousands of years. It is important to empower

> and respect the local leaders. I think they play a powerful role in terms of communicating climate changes, environmental issues and natural disasters. Despite the latest technology, never underestimate the power of the leaders in the community – the chiefs, youth leaders, the women leaders, the church leaders and pastors and the network they have. They have a lot more influence than even the political leaders, the business leaders.
>
> (F Samuel, 2017, personal communication, 27 October)

Cross-sectoral collaboration

This conversation highlights the importance of cross-sectoral partnership. After the cyclone a quick response of communication specialists from the private sector working alongside the public sector assisted in providing emergency communication.

Various regional organisations work with the Vanuatu government and civil society in capacity building so that this most risk-prone country in the world can meet the challenges posed by climate change and other natural disasters. An example of this is the Vanuatu NGO Climate Change Adaptation Programme whereby different agencies and organisations work under a unifying framework in climate adaptation and disaster risk reduction at community level. The Programme consists of a consortium of six local and international agencies[8] which work with communities using the resilience framework especially developed for Vanuatu (Oxfam, 2015).

The members of the consortium work to their own strengths but share their best practice, expertise and methodologies. This model, based on networking, collaboration, facilitation and community empowerment, means while adding value to existing responses to climate change they do not compete for the same funds. The government does not have the capacity in all these areas, but provides policy guidelines and support through the Vanuatu National Advisory Board on Climate Change and Disaster Risk Reduction. Resilience and climate change education is embedded in all government policies and K-12 school curricula, as well as in vocational programmes.

For Vanuatu, food and water security are prime climate change issues while sea level rise and erosion are down the list of priorities. Christopher Bartlett from the German Agency for International Cooperation (GIZ) helps me to understand how improving agricultural practice ensures food security and builds resilience. The ability to grow food allows people not only to feed themselves, but also to generate an income which pays for children's school fees and other necessities.

Food security means improving agriculture, forestry, fisheries and livestock practices. Access to ground water is limited in Vanuatu and many communities are completely dependent on rain. A prolonged drought can be debilitating for communities which rely solely on farming for their income. Transformation to resilience prepares communities to develop different sources of income instead of being reliant only on subsistence farming or planting only one type of crop. Expanding into cottage industries such as virgin coconut oil production, or having a poultry farm, adds to diversity of income sources.

Traditional knowledge and weather forecasts

In indigenous epistemology, nature, culture and communication are inextricably interlinked in creation and preservation of the places where humans live, the things we eat and ways in which we safeguard ecological knowledge for the next generation. Forsyth (2013: 1) describes traditional knowledge as comprising "traditional cultural expressions – such as songs, stories, oral traditions, visual and performing arts, and ritual and cultural practices – as well as biological knowledge, innovations, and practices". Traditional ecological knowledge (TEK) refers to the way in which indigenous cultures understand and relate to local ecosystems (Lewis, 1993). The TEK system is based on people's close observation of nature and the natural phenomenon in which humans are as much part of the ecological systems as non-humans (Adger, 2003).

The importance of place and collective community input means that the nature culture nexus is deep rooted and informs all aspects of the indigenous world. In documenting indigenous knowledge and epistemology of her clan, Fijian scholar, Unaisi Nabobo-Baba (2006: 42) employs the *Vanua* research method to show how indigenous Fijians perceive the world:

> Amongst the Vugalei Fijians the world is one big entity and all things in it are associated. Life is complete and wholesome when all elements – air, wind, seas, rivers, plants, animals, fishes, people, the dead – are 'synchronised'. The Vugalei do not find it easy to distinguish land from people, or people from their clans, villagers, kinship and religion.

Communities and meteorologists in five Pacific island countries, Vanuatu, Niue, Samoa, Solomon Islands and Tonga, are engaged in gathering information about traditional forms of weather forecasts as part of a Regional Climate Traditional Knowledge project (Walshe et al., 2017; Percival, 2008; SPREP, 2016, 2017). Working in close cooperation with communities, the department of meteorology in each country combines weather forecasts with traditional knowledge to build community resilience to extreme events. The regional initiative began in 2016 to include traditional knowledge with the scientific knowledge used by meteorologists. It is supported by the Australian Bureau of Meteorology and the Secretariat of the Pacific Regional Environment Programme (SPREP), an intergovernmental organisation, which assists Pacific islands governments on climate change and environmental issues.

The work includes collecting, documenting, storing and analysing traditional knowledge stories to identify climate indicators used by pilot communities. Of the 100 indicators that were collected from around the country in Vanuatu, about six were well aligned. If the chief from a village calls to report certain fish behaviour, and it is one of the aligned indicators, then the Department of Meteorology will include the traditional knowledge in the forecast warning of a rain event.

Bridging the science and TK divide

Siosinamele Lui, the Climate Traditional Knowledge Officer at SPREP, explains the significance of the partnership in this interview:

> What I've found really interesting about this project is that it makes it easier to have conversations about climate change/climate science with communities. The TK project is a communication tool, and it bridges the gap between science and social science, between policies and technical work in the world of meteorological services. What may be difficult to explain in meteorological terms can be easily understood through traditional knowledge because that is what our Pacific communities can relate to and understand. When talking about seasonal forecasting, not everyone can understand the technicality of the information, but if we use traditional knowledge, the seasonal terminology, language and prediction techniques communities are used to, it increases the intake of scientific information as well as increasing community resilience. In some communities, they know that it will be an active cyclone season if the turtles nest higher up in the sand dunes (Vanuatu TK). This knowledge adds value to seasonal forecasts.
>
> From looking at the biological (animal and plant behaviour) and astronomical (moon phases, stars) indicators our Pacific communities knew what will happen two to three months ahead

> before the onset of the El Niño Southern Oscillation (ENSO) phases.
>
> Traditional knowledge provides the extra information, a traditional warning system which allows communities to prepare, respond and utilise traditional coping strategies and mechanisms that have increased communities' resilience.
>
> (SPREP, 2017)

Traditional Knowledge is a sensitive area because communities are protective of their knowledge. The project has developed intellectual property guidelines on what is and is not acceptable in discussing climate knowledge. There are several levels of knowledge: widely known, to most secret *tabu* knowledge. The programme uses knowledge that can easily be transferred to another group to help them solve some of their climate problems, while the community that shares the knowledge gets acknowledgement.

Faith-based organisations

People in the Pacific are predominantly Christian. This places a responsibility on churches to communicate the growing threat of climate change to their constituencies. Belief in Christian narratives, such as God's promise to Noah to never flood the earth, often complicates public awareness. Some people demonstrate blind faith in terms of divine protection from natural disasters. This creates barriers to communication of knowledge of emergency preparedness, due to conflict between disaster information and religious beliefs (Tacchi et al., 2013a).

How has this belief system become so entrenched in Pacific people's faith and their cultural practice, and how can it be changed? Rev. Seforosa Carroll, Manager, Church Partnerships Pacific at Uniting-World explains:

Where is god in this?

> There is internal work that needs to be done in the Pacific and that is to do with the theology and recovery, I believe, of Pacific island people's indigenous epistemologies. Their ancient wisdom is something that needs to be reconnected with and tapped into to help them wrestle with and engage with the challenge of climate change. The barrier and challenge is that because of the onset of colonisation and Christian mission the very thing

> that they need for their survival, their indigenous epistemology, has become something that they have been taught to treat with suspicion. So, we have a task there already about reconnecting that worldview and that spirituality. In the Pacific, since time immemorial, we had a holistic world view where people were connected to land and sea and community.
>
> There are a number of critical theological issues in the Pacific concerning climate change such as theodicy, the problem of evil, and the question of where God is in our suffering. Where is God? A response to this question is a deep belief that God will save and we just need to faithfully wait. This understanding actually leads to disengagement with reality in the here and now. I don't say this with judgement because I do understand that when you have no resources to get out of your island that is disappearing, you have to go back to faith, because it is your only anchor. What else do you have? So, for people mainly in Tuvalu and Kiribati, there is the understanding that Noah's Ark will come again; we just have to faithfully wait. The emotional and psychological burden can become so great, it becomes difficult for people to see beyond the present and to a more hopeful future.
>
> The other challenge is stewardship. How do we as a community and people care for the earth the way we are called to? What we hope to do is to help people to engage and to see it as part of their spirituality. The pulpit is a critical medium of teaching and imparting faith. It is a powerful medium through which to empower people to engage. But there is a need to understand our relationship with God and Creation differently than what we have been taught unhelpfully to believe. This would require changes in the theological curriculum, and how pastors and ministers are trained and formed for ministry. You've got to start from the grassroots; you've got to start with things as simple as Sunday school. These are the kinds of ways we hope to get some sort of change. What you start today will take at least another 20 years, another generation to train to bring about a new way of thinking.
>
> (Carroll, 2017)

Faith-based networks

The Pacific Conference of Churches (PCC), a regional ecumenical organisation of 29 member churches from around the Pacific, uses its strong network of church leaders, and women's and youth organisations to prepare communities for the impact of climate change. Special emphasis is placed on educating women's groups and church leaders who use their horizontal social networks to take the message more deeply into the community using interpersonal communication (Tacchi et al., 2013b).

PCC also employs its vertical networks, working alongside government agencies, to support communities during the relocation phase as villagers are uprooted from their coastal homes and moved to higher ground. Member churches are trained to offer psychological trauma healing to communities who have been displaced from their ancestral homes. This includes workshops which explain the science of climate change to help people have a better grasp of the changes in their natural environment and the new challenges to social cohesion as communities move to locations which are closer to transport and mobile networks.

In Australia, Pacific Calling Partnership (PCP), an initiative of the Edmund Rice Centre, a Catholic NGO, creates spaces for I-Kiribati and Tuvaluan voices in international forums. PCP prepares the next generation of Pacific leaders by training them in advocacy and community education. The aim is to raise awareness of the situation in these countries by facilitating people-to-people exchange with politicians, policy makers, private industry and community groups. Director, Phil Glendenning explains:

> Face-to-face communication is vital. Kiribati is a line in the sand on this issue. The aim is that by meeting a flesh and blood human being who has lived the experience of this issue and is sitting right in front of you, it becomes much harder to turn away. Moreover, the profoundly important thing to do is to get the people who are most affected to have a stake in this, to have a say in this and be heard. It is far more effective and far more significant when those with the lived experience engage with those who are sceptical.
>
> (P Glendenning, 2017, personal communication, 23 October)

Groups of young people from Kiribati and Tuvalu visit Australia to share stories about how the changing climate is affecting their families and their communities, thereby joining the dots between the broader global debates about climate change and what is occurring at a personal level.

Churches are an important node in the communication network. Christian teachings are deeply embedded in the daily practices of Pacific island communities. The examples in this chapter demonstrate that churches play a critical role in filling the knowledge gap and assisting communities to transition from a closed mindset to an educated understanding of climate change and natural disasters. Messages need to be grounded in scientific facts alongside traditional knowledge, yet woven in a language that does not undermine people's faith in God.

Conclusion

My journey in the Pacific and conversations with diverse individuals have highlighted two important lessons. First is the value of interpersonal dialogue, and second is the importance of networks in social cohesion. They are both inextricably linked in resilience-building.

Pacific culture, language and stories are based on oral tradition. Talanoa fosters informed knowledge exchange through trusted social networks. These stories are place-based and contain local knowledge and history – information that cannot be gathered through purely quantitative measurement (Harper et al., 2012: 96; Janif et al., 2016). In summarising the features of place-based storytelling and listening, Matthews and Sunderland (2017: 145–146) remind us that:

> embodied and emplaced listening potentially creates a more deeply 'contextualised' view of storytellers and stories among listeners. It can also create opportunities for listeners to directly sense, intuit, feel, see, hear, touch, taste, or smell what the storyteller might have experienced and portrayed in their story.

By telling their own stories, community actors are able to "call into question the authority and legitimacy of established storytelling institutions" (Spurgeon, 2015: 137). The narrative form connects people in a meaningful way – "face to face, eye to eye, mind to mind, heart to heart" is the way Storytellers NSW describes the art of storytelling.[9] These encounters are the essential first step in engaging people in communication about their own circumstances.

Storytelling uses a dialogical approach and values the agency of subaltern groups. It is transformative, enumerates the problem close-up by highlighting the struggles and adversities, and defines one's identity. Storytelling is one of the most effective forms of interpersonal communication, and one that should be included in the participatory environmental communication toolbox.

Notes

1 https://cop23.com.fj/fiji-clearing-house-risk-transfer-designed-build-greater-resilience-vulnerable-nations-cop23-president

2 Talanoa sessions, facilitated by Halapua, were used in Fiji to reconcile differences between the political parties and mend ethnic divisions in an effort to bring about national unity after the 2000 coup.
3 Not all interviews followed the talanoa protocol. Interviews with representatives from government, NGO and faith-based organisations were conducted during informal interview sessions.
4 His Excellency Anote Tong Address – Pacific Climate Change Conference Victoria University, Wellington, New Zealand, 15–17 February 2016.
5 https://kirican.wordpress.com/about-kirican/
6 "A theory of distributive justice is a central feature of any fully worked out ethical system, but, as will become clear, it is especially important in the debate about the impacts of climate change" (Page, 2006: 3).
7 Fred Samuel is currently serving as Chief Executive Officer and Director of Pacific Group Limited in Vanuatu.
8 The consortium includes CARE International, Oxfam, Save the Children, Vanuatu Red Cross, Vanuatu Rural Development Training Centres Association and the Deutsche Gesellschaft für Internationale Zusammenarbeit (GIZ) [German Society for International Cooperation].
9 www.storytellersnsw.org.au/

References

Abate, R. S. 2010. Public nuisance suits for the climate justice movement: The right thing and the right time. *Washington Law Review*, 85: 197–252.

Adger, W. N. 2003. Social capital, collective action, and adaptation to climate change. *Economic Geography*, 79(4): 387–404. doi:10.1111/j.1944-8287.2003.tb00220.x

Anderegg, W. R., Prall, J. W., Harold, J. & Schneider, S. H. 2010. Expert credibility in climate change. *Proceedings of the National Academy of Sciences*, 107(27): 12107–12109. doi:10.1073/pnas.1003187107

Anderson, I. 2013. *The Economic Costs of Noncommunicable Diseases in the Pacific Islands: A Rapid Stocktake of the Situation in Samoa, Tonga, and Vanuatu* [Online]. Washington, DC: The World Bank. Retrieved from http://documents.worldbank.org/curated/en/291471468063255184/pdf/865220WP0Econo0Box385176B000PUBLIC0.pdf

Bündnis Entwicklung Hilft. 2017. *WorldRiskReport: Analysis and Prospects 2017* [Online]. Berlin: Bündnis Entwicklung Hilft. Retrieved from https://reliefweb.int/sites/reliefweb.int/files/resources/WRR_2017_E2.pdf

Caritas Oceania. 2017. *Turning the Tide: Caritas State of the Environment for Oceania 2017 Report* [Online]. Thorndon, Wellington: Caritas Aotearoa New Zealand. Retrieved from https://reliefweb.int/sites/reliefweb.int/files/resources/State%20of%20the%20Envrionment%20for%20Oceania%202017%20report_0.pdf

Carroll, S. 2017. *Pacific Stories – Courageously Encountering God in Climate Change*. Edmund Rice Centre, Homebush West, NSW: Pacific Calling Partnership.

Cook, J., Nuccitelli, D., Green, S. A., Richardson, M., Winkler, B., Painting, R., Way, R., Jacobs, P. & Skuce, A. 2013. Quantifying the consensus on anthropogenic global warming in the scientific literature. *Environmental Research Letters*, 8(024024): 1–7. doi:10.1088/1748-9326/8/2/024024

Farbotko, C. 2018. Voluntary immobility: Indigenous voices in the Pacific. *Forced Migration Review*: 57.

Forsyth, M. 2013. How can traditional knowledge best be regulated? Comparing a proprietary rights approach with a regulatory toolbox approach. *Contemporary Pacific*, 25(1): 1–31. doi:10.1353/cp.2013.0004

Freire, P. 1970. *Pedagogy of the Oppressed*. New York: Herder and Herder.

Green Climate Fund. 2017. *About the Fund* [Online]. Yeonsu-gu, Incheon The Green Climate Fund. Retrieved 19 November 2017 from www.greenclimate.fund/who-we-are/about-the-fund

Harper, S. L., Edge, V. L. & Willox, A. C. 2012. "Changing climate, changing health, changing stories project": Using an ecohealth approach to explore impacts of climate change on Inuit health. *EcoHealth*, 9(1): 89–101. doi:10.1007/s10393-012-0762-x

Hewitt, C., Mason, S. & Walland, D. 2012. The global framework for climate services. *Nature Climate Change*, 2(12): 831–832. doi:10.1038/nclimate1745

Janif, S. Z., Nunn, P. D., Geraghty, P., Aalbersberg, W., Thomas, F. R. & Camailakeba, M. 2016. Value of traditional oral narratives in building climate-change resilience: Insights from rural communities in Fiji. *Ecology and Society*, 21(2): 7. doi:10.5751/ES-08100-210207

Klein, N. 2014. *This Changes Everything: Capitalism vs. the Climate*. New York: Simon & Schuster.

Lewis, H. T. 1993. Traditional ecological knowledge: Some definitions. *In:* Williams, N. M. & Baines, G. (eds.) *Traditional Ecological Knowledge: Wisdon for Sustainable Development*. Canberra: Centre for Resource and Environmental Studies, ANU.

Matthews, N. & Sunderland, N. 2017. *Digital Storytelling in Health and Social Policy: Listening to Marginalised Voices*. London: Routledge.

Nabobo-Baba, U. 2006. *Knowing and Learning: An Indigenous Fijian Approach*. Suva: Institute of Pacific Studies, University of the South Pacific.

Oreskes, N. 2018. The scientific consensus on climate change: How do we know we're not wrong? *In:* Lloyd, E. A. & Winsberg, E. (eds.) *Climate Modelling: Philosophical and Conceptual Issues*. Cham, Switzerland: Springer.

Oxfam. 2015. *Lessons from the Vanuatu NGO Climate Change Adaptation Program* [Online]. Carlton, VIC: Oxfam. Retrieved from https://www.oxfam.org.au/lessons-from-the-vanuatu-ngo-climate-change-adaptation-program/

Paeniu, L., Iese, V., Jacot des Combes, H., De Ramon, N. Y., Korovulavula, I. T., Koroi, A., Sharma, P., Hobgood, N., Chung, K. & Devi, A. 2015. *Coastal Protection: Best Practices in the Pacific* [Online]. Suva, Fiji: Pacific Centre for Environment and Sustainable Development. (PaCE-SD). The University of the South Pacific. Retrieved from http://repository.usp.ac.fj/8219/1/Paeniu_et_al._2015_Best_Practices_in_Coastal_Protection_for_the_Pacific.pdf

Page, E. A. 2006. *Climate Change, Justice and Future Generations*. Glos, UK: Edward Elgar Publishing Limited.

Percival, G. S. 2008. *An Assessment of Indigenous Environmental Knowledge (IEK) in the Pacific Region to Improve Resilience Environmental Change* [Online]. NSW, Australia: The Climate Change Research Centre, University of New South Wales. Retrieved from http://www.academia.edu/12621790/An_Assessment_of_Indigenous_Environmental_Knowledge_IEK_in_the_Pacific_Region_to_Improve_Resilience_to_Environmental_Change

Pettit, J. 2004. Climate justice: A new social movement for atmospheric rights. *IDS Bulletin*, 35(3): 102–106. doi:10.1111/j.1759-5436.2004.tb00142.x

ReliefWeb. 2018. *Pacific: Drought – 2015–2017* [Online]. Papua New Guinea: ReliefWeb. Retrieved from https://reliefweb.int/disaster/dr-2015-000127-fji

Robinson, D. & Robinson, K. 2005. *"Pacific ways" of Talk: Hui and Talanoa* [Online]. Wellington, New Zealand: New Zealand Trade Consortium Working Paper, No. 36. Retrieved from www.econstor.eu/bitstream/10419/66100/1/494642475.pdf

Schalatek, L. 2017. Green Climate Fund: Signalling Effect for Global Climate Finance. *D+C Development and Cooperation*, 21 September 2017. Retrieved from www.dandc.eu/en/article/green-climate-fund-important-implementing-paris-agreement

Sen, A. 1999. The ends and means of development. *In: Development as Freedom*. Oxford: Oxford University Press.

Sen, A. 2013. The ends and means of sustainability. *Journal of Human Development and Capabilities*, 14(1): 6–20. doi:10.1080/19452829.2012.747492

SGI Quarterly. 2007. Talanoa-talking from the heart: Interview with Dr. Sitiveni Halapua. *SGI Quarterly*: 47. Retrieved from www.sgi.org/content/files/resources/sgi-quarterly-magazine/0701_47.pdf

SPC. 2017. *Pacific Community Results Report 2016* [Online]. Noumea, New Caledonia: Pacific Community (SPC). Retrieved from www.spc.int/wp-content/uploads/2017/07/CRGA-47-Results-Report-2016.pdf

SPREP. 2016. *Pacific Islands Meteorological Services in Action: A Compendium of Climate Services Case Studies* [Online]. Apia, Samoa: Secretariat of the Pacific Regional Environment Programme (SPREP), WMO and Environment and Climate Change Canada. Retrieved from https://library.wmo.int/opac/doc_num.php?explnum_id=3394

SPREP. 2017. *Q and A: The Role of Traditional Knowledge in Pacific Meteorology with Siosina Lui of SPREP* [Online]. Apia, Samoa: Secretariat of the Pacific Regional Environment Programme (SPREP). Retrieved from 30 December 2017 www.sprep.org/climate-change/q-and-a-the-role-of-traditional-knowledge-in-pacific-meteorology-with-siosina-lui-of-sprep

Spurgeon, C. 2015. Participatory media and "co-creative" storytelling. *Media International Australia*, 154: 132–138. doi:10.1177/1329878X1515400116

Tacchi, J., Horst, H., Papoutsaki, E., Thomas, V. & Eggins, E. 2013a. *PACMAS State of Media and Communication Regional Report 2013* [Online]. Melbourne, Australia: PACMAS/ABC International Development. Retrieved from www.pacmas.org/profile/pacmas-state-of-media-andcommunication-report-2013/

Tacchi, J., Horst, H., Papoutsaki, E., Thomas, V. & Eggins, E. 2013b. *PACMAS, Fiji State of Media and Communication Regional Report 2013*. [Online]. Melbourne, Australia: PACMAS/ABC International Development.

The Movement for Community-Led Development. 2017. *Saemaul Undong – the Republic of Korea's New Village Movement, Part 1* [Online]. Washington, DC: The Movement for Community-Led Development. Retrieved from https://communityleddev.org/2016/03/29/saemaul-undong-the-republic-of-koreas-new-village-movement-part-1/

Tong, A. 2016. *His Excellency Anote Tong Address. Pacific Climate Change Conference*. 15–17 February 2016, Victoria University, Wellington, New Zealand.

UNISDR. 2017. *Can COP23 Fill the Ambition Gap?* [Online]. Geneva, Switzerland: United Nations Office for Disaster Risk Reduction. Retrieved 29 December 2017 from www.unisdr.org/archive/55932

Walshe, R. A., Chang Seng, D., Bumpus, A. & Auffray, J. 2017. Perceptions of adaptation, resilience and climate knowledge in the Pacific: The cases of Samoa, Fiji and Vanuatu. *International Journal of Climate Change Strategies and Management*. 10(2): 303–322 doi:10.1108/IJCCSM-03-2017-0060

WCED. 1987. *Our Common Future (Brundtland Report)*. Oxford: United Nations World Commission on Environment and Development, Oxford University Press.

5 Engaging communities in environmental communication

Reflection

On 16 July 1945 the first atomic bomb was detonated successfully in the desert of New Mexico in the United States. Little did people living in the pristine Pacific Ocean realise that event would become their source of grief. In the ensuing half-century hundreds of nuclear bombs were tested by colonial powers which had jurisdiction over various Pacific islands. The inhabitants of these islands were powerless to drive the occupiers from their own land. But Pacific islanders did have the power to network, organise and protest against the environmental injustices being visited upon them. I was a young journalist working for the only Fijian daily newspaper, The Fiji Times, when the anti-nuclear protest movement began gaining strength in the late 1970s. Even in a small country like Fiji the wind of change was blowing. I found myself in the middle of two movements – the anti-nuclear movement and the women's movement. With assignments to cover the marches and attend rallies, I had the good fortune to meet activists with indomitable courage. Some of the most outspoken amongst them were women who were finding their own voice by defining the struggles of their people. Women like Amelia Rokotuivuna, Darlene Keju and Suliana Siwatibau were, and remain, awe-inspiring women of the Pacific, who fought to change the status quo. In the book co-authored by Suliana Siwatibau and David Williams during the height of the anti-nuclear movement in the 1980s, they make an appeal, "We, the ordinary people of the world, must do something to make sure that we are not heading for the destruction of ourselves and our earth.

And we must do something quickly, before it is too late" (1982: 6). The passion and courage of these women activists informed my own political growth during those formative years and challenged me to reimagine my own future in the wider world.

Introduction

The story of modern environmental movements is one of people power. When people aspire for change they will take action to bring it about. As communities around the world become aware of large-scale degradation of their waterways, forests, oceans and air they turn to civic action and mass protest to demand change. From the streets of New York to the Himalayan villages in northern India, from the forests of Tasmania to the coral lagoons of the Pacific islands, citizens band together to protest against chemical poisoning, deforestation and nuclear testing in their 'backyard'. Concerns about preserving the ecological balance, inequitable distribution of resources, social justice, and emancipation of marginalised and displaced peoples mobilise grassroots action for social change. The discontent felt by individuals and communities in their everyday lives progressively acquires broader 'networks of solidarity' as local direct action is supported by national and global acts of resistance (Dutta, 2012: 34).

Environmental movements are born out of deep ecological concern in ordinary people about threats to their natural habitat resulting from unsustainable development activity. At the heart of such protest movements is an agenda for social, economic and political change. That agenda seeks a better deal for both the natural and human environment through policy reforms and promulgation of social values that ensure sustainable use of natural resources. The ecological impact of human action has led to air and water pollution, deforestation, loss of biodiversity, danger from radioactive waste disposal, long-term climate change impacts such as desertification, rising sea levels and frequent extreme weather events, to name just a few.

Expanding on the range of the 1970s and '80s environmental movement, contemporary environmental protests seek to protect traditional sites, and conserve forests, rivers and land for future generations. Environmental activists have challenged multinational interests whose sole purpose is to maximise profit, and governments focused on national economic growth at the exclusion of all else. Examples of these include the Chipko Movement in India, Franklin River Dam protests in Tasmania, The Dakota Access Pipeline protests by Standing Rock Sioux Tribe in the United States, and battles between mining giants and traditional communities in the Pacific.

This chapter reviews key milestones in the modern environmental movement and the role of communication networks in three different protest movements. It then charts the emergence of the anti-nuclear movement in the Pacific region as a consequence of 50 years of nuclear testing. The chapter explores the significance of diversity, network and agency in the broader context of global social movements and regional alliances. Links are drawn to networks of solidarity that have given Pacific peoples a common voice against neo-colonial policies, resource exploitation by foreign powers and a desire for political independence.

Global environmental awareness

The modern-day environmental movement is a manifestation of people's rising concern for the natural world and a call for better stewardship of planet Earth. Between 1800 and the mid-1900s countries in the northern hemisphere underwent rapid industrialisation as fossil fuels in the form of oil and coal provided a cheap and abundant source of energy to power the industrial complex. While economic growth and technological innovation brought many benefits such as electrification, better yields in agricultural food production and greater affordability in car and home ownership, the use of coal power has been responsible for many ills, both to human health and the natural environment. People in industrialising countries suffered from the effects of carbon pollution as millions of tonnes of coal were burned to power mills and factories. One of the worst examples of carbon pollution is the Great Smog of 1952 in the United Kingdom. It took place during an unusually cold period in December when toxic smoke from factory chimneys was trapped in the lower atmosphere as a result of an anti-cyclone in which a layer of warm air in the higher atmosphere trapped the cold air below. The UK Met Office describes its effects:

> A fog so thick and polluted it left thousands dead wreaked havoc on London in 1952. The smoke-like pollution was so toxic it was even reported to have choked cows to death in the fields. It was so thick it brought road, air and rail transport to a virtual standstill.
> (The Great Smog of 1952, 2015)

Impurities in the atmosphere were measured in tonnes. They included dangerous levels of carbon dioxide, hydrochloric acid, fluorine compounds, and sulphur dioxide which had converted into sulphuric acid. More than 4,000 people died as a direct result of the Great Smog, but that number could be increased three-fold because of associated ailments. Parallels can be drawn today with the industrialising countries of China, India and Indonesia.

Evidence of man-made environmental degradation emerged in other places in the northern hemisphere throughout the 1960s. In Sweden, there was concern about pollution in the form of sulphur emissions from the United Kingdom and Central Europe. The resulting acid rain and the acidity of Scandinavia's lakes and streams caused large-scale fish death. The health and productivity of forests' soils and trees were also damaged. This prompted Sweden to enact environmental regulations in 1967, becoming the first country in the world to do so (Johnson, 2012). Sweden also proposed the idea of a United Nations conference to look at the impact of human activity on the environment. In putting the case for the conference, Sweden's Ambassador to the UN, Sverker Astrom, urged world leaders to find solutions to environmental problems since it was in the interest of all peoples on earth, because "Man depends for his survival on an infinitely complex system of relationship and balance between innumerable living organisms, all existing in or on the extremely thin crust of earth or just above it" (Johnson, 2012: 9).

The United Nations Conference on the Human Environment was held in Stockholm in 1972, bringing together countries of the Global South and Global North in a dialogue which looked beyond environmental problems to issues of poverty and inequitable distribution of wealth, identified as the key causes of environmental degradation. The conference established the United Nations Environment Programme (UNEP) with a remit to provide leadership and guide environmental activities within the UN framework. In 1983 the UN set up the Brundtland Commission to produce a common global vision for sustainable development. The commission's report titled *Our Common Future* (1987), also referred to as the Brundtland Report, defined sustainable development as "development that meets the needs of the present without compromising the ability of future generations to meet their needs" (WCED, 1987: 44). This definition continues to inform current discourse on environmental sustainability and the United Nations Sustainable Development Goals.

Modern environmentalism has strong links to the environmental activism of the Sierra Club in the United States and to the 1970s grassroots movement arising out of ordinary people's concern for their environment. Across the United States, great natural spaces were 'tamed' by emptying lakes and filling in flowing creeks to build roads and cities during the industrial revolution. Ten years after the Great Smog in Britain, American marine biologist and nature writer, Rachel Carson, warned against the hazards of chemical pesticides, namely Dichlorodiphenyltrichloroethane (DDT). In her critically acclaimed book *Silent Spring*, Carson (1962: 5) questioned humans' domination of nature made possible by modern scientific discoveries: "Only within the moment of time represented by the present century has one species – man – acquired significant power to alter the nature of the world". Her book depicted a future world where spring no longer meant a joyous celebration of life because all life – birds, apple blossoms, fish and even children – had been 'silenced' by the insidious effects of DDT. Carson warned of ecological destruction of whole natural systems caused by humans' attempts to control pests, and through the industrial manufacture and indiscriminate use of DDT. Her singular attempts to expose the devastating role of man-made pesticides raised public awareness about the fragile nature of the environment and the intricate interconnectedness of the ecosystem.

Concerns about oil spills, nuclear testing and the use of insecticides catalysed the agency of people culminating in the first Earth Day held on 22 April 1970 in the United States. The people-led environmental action brought 20 million Americans into the city streets, parks, town squares and foreshores all across the United States. Students, housewives, labourers, farmers and office workers converged in public spaces demanding action from their government to become better guardians of the land, rivers, lakes, air and all living things. Founder of Earth Day, Senator Gaylord Nelson, recalled the immediate and immense response to his announcement for Earth Day:

> Earth Day worked because of the spontaneous response at the grassroots level. We had neither the time nor resources to organize 20 million demonstrators and the thousands of schools and local communities that participated. That was the remarkable thing about Earth Day. It organized itself.
>
> (Nelson Institute for Environmental Studies, 2016)

Telegrams, letters and telephones were used to network and organise local action. Diverse groups that had fought against oil spills, polluting factories and power plants, raw sewage, toxic dumps, pesticides, freeways, the loss of wilderness and the extinction of wildlife, and nuclear testing suddenly realised they shared common values and joined forces (The History of Earth Day, 2016). An environmental movement was born that would become global.

In the Global South, the structural inequities in the global economic system controlled by imperial powers framed the debates around environmentalism aligned with democratic and human rights principles. Independence from colonial rule would emancipate developing nations and give them greater control over land and water resources. Examples include the Gandhian independence movement of 'Satyagraha' an action of non-violent non-cooperation against the British rule in India that demanded the right of citizens to have free access to natural resources, and Julius Nyerere's 'Ujamaa' policy in Tanzania that gave greater importance to family and community in social and economic development (Chabot and Vinthagen, 2007; Ibhawoh and Dibua, 2003). While these forms of protest were not aimed at environmental protection per se, they drew important links between the independence movement and the rights of local communities to access natural resources for their own livelihood.

Environmental activism as participatory networks

Successful social movements require the participation of ordinary people. These movements gain momentum at ground level relying on the resources of trusted social networks. Castells (2009: 300) says that social movements are a result of 'social actors aiming for cultural change' which he explains as "a change of values and beliefs processed in the human mind on a scale large enough to affect society as a whole". Political change occurs when institutions adopt these new values in response to cultural change. Effective communication and collaboration among actors are essential elements in achieving this transformation, as Castells (2009: 301 italics in original text) elucidates:

> Social movements are formed by communicating messages of rage and hope. The specific structure of communication of a given society largely shapes social movements. In other words, social movements, and politics, insurgent or not, spring up and live in the *public space*. Public space is *the space of societal, meaningful interaction where ideas and values are formed, conveyed, supported, and resisted; space that ultimately becomes a training ground for action and reaction.*
>
> (author's italics)

Castell's use of the term 'public space' aligns with the Habermasian public sphere where the ferment of ideas and political action takes place. The public sphere is an important concept in social movements, as Brulle (2010: 4) explains: "The public sphere is where social movement organizations can identify problems, develop possible solutions, and create sufficient political pressure to have them addressed by constitutional governments". German philosopher Jürgen Habermas developed the concept of public sphere which is "made up of private people gathered together as a public and articulating the needs of society with the state" (Habermas, 1989: 176).

Bottom-up channels of communication play an integral role in social movements as part of the broader public sphere. An essential feature of social movements is the use of networks through which information is disseminated and momentum is built. Communication is necessary to pool resources and to organise and coordinate the efforts of individual members and civil society organisations towards achieving collective action (Krinsky and Crossley, 2014).

Scholars who have studied social movements agree that participation of people through grassroots networks is an important process of mobilising citizens (Milan, 2013; Brulle, 2010; Downing, 2001). Both interpersonal and mass mediated communication networks are needed to share messages of solidarity and resistance. CIVICUS, which represents civil society organisations (CSO) globally, provides an insight into such movements: "What mass protests have in common is that they signify anger and frustration. People are angry at not having their voices heard, and they feel excluded and humiliated. They are demanding their right to a voice, and seeking accountability" (CIVICUS, 2016: 5). These civil society networks enlist mediated communication such as social media and alternative media to expand their reach and influence in an effort to mobilise citizens to seek change.

Citizens need safe spaces from which to voice their concerns as public spaces are being diminished. Increasingly these are being replaced by alternative safe spaces in the digital sphere in societies with coercive regimes and punitive anti-protest laws against citizens such as the case in Australia where there are increased penalties against environmental groups which oppose coal seam gas sites (Kretser, 2016).

In recent times, community action has been replaced largely by strategic communication campaigns which privilege elite voices. While the environmental movement was born out of grassroots activism, progressively that voice has been taken over by powerful global environmental non-governmental organisations (ENGO) which claim to speak on behalf of ordinary people. Public rage has been outsourced to communication experts who produce professional strategic messages for well-funded ENGOs that use the power of celebrity to garner public support. While ENGOs are an important part of civil society, their reliance on professional top-down strategic communication models has disenfranchised citizen agency and disempowered direct community action. This proposition was well articulated in Brulle's (2010: 82) concerns about 'environmental identity campaigns' which apply cognitive science and psychological approaches in message design. These, he argues, are elite-driven social changes and fail to provoke conversation and debate. They use discursive frames which are situated within hegemonic worldviews and as such are "embedded in power relationships, and authorize certain actors and perspectives while neglecting others" (Brulle, 2010:85). He states:

> Thus, while identity campaigns may offer short-term advantages, they are most likely incapable of developing the large-scale mobilization necessary to enact the massive social and economic changes necessary to address global warming.
>
> (2010: 83)

In Brulle's view, critical communities and social movements play an important role in the creation and advocacy of alternative field frames which challenge the hegemonic worldview. They mobilise citizen participation, and as such "a participatory structure is a key component in large-scale social change efforts" (Brulle, 2010: 84).

Use of communication in environmental protest movements, both past and present, provides some interesting insights into how communication processes assist citizen mobilisation, their use of support networks, and the key elements necessary for the success of these movements. Participatory communication provides an important platform for local communities to share knowledge and create awareness about local environmental issues through their own unique perspectives. The expansion of this form of public sphere, both for dialogue and political action, has helped to engage disenfranchised communities such as women, youth, people with disabilities and indigenous communities on critical issues such as climate change.

The next section looks at the use of networks and participatory communication strategies applied in three iconic environmental movements. The 1970s Chipko Movement in India was an act of embodiment of nature by tribal communities; the 1980s Tasmanian dam protest in Australia employed interpersonal communication and analogue media to gather support; and the Standing Rock Sioux against the Dakota Pipeline in 2016 used social media to enlist a diverse support network.

The Chipko Movement – agency

The Chipko Movement in India was ground-breaking in the history of the environmental protest movement. The protest movement is mainly known for the agency enacted by women forest dwellers of the sub-Himalayan region who embraced the trees in order to protect them from commercial exploitation in 1974. The hill people of the Garwahl district maintained the traditional eco-balance while using local resources for food and shelter. Jain describes the chain of events which led the women to form a human chain around the trees when timber company workers turned up to cut down the trees in the Reni forest:

> Gaura Devi described the encounter in graphic detail by commenting on the rude behavior of some men and her courage in pushing herself forward in front of the gun of one of the labourers. She challenged the man to first shoot her down, then only to touch the trees. She compared the forest with her mother's home (maika). By facing the men squarely, she and her companions forced them to retreat.
>
> (Jain, 1984: 1792)

The Chipko Movement is an example of a people's deep cultural connection to its natural environment. Chipko, meaning to embrace or hug, embeds the notion of human-nature solidarity. By hugging the trees, the women enacted their agency in a spontaneous act of protecting their "traditional harmonious relationship with nature" (Chatterjee, 2012: 11). They understood the intrinsic value of the forest in the maintenance of their ecosystem and as a source of their livelihood. As an act of defiance they used their own physical form to communicate their message in a desperate attempt to save their forests from short-term commercial exploitation.

Bandyopadhyay (1999) argues that the Chipko Movement was the result of many years of grassroots organising by Gandhian activists who dispersed to rural India after independence, teaching communities about sustainable livelihoods and forest protection. Women were especially encouraged to attend these knowledge gatherings and intrinsically understood the need for forest protection to achieve continued sustainable livelihoods, since they were the main food gatherers for the family. As head of the local women's organisation, Gaura Devi was able to call upon her network of 30 women and their children to form the ring of protection. The women may have formed the idea of embracing the tree from a song written by folk poet Ghanashyam Raturi:

> Embrace the trees in the forests
> And save them from being felled!
> Save the treasure of our mountains
> From being looted away from us!!
> (Bandyopadhyay, 1999: 880)

The concerted actions of different grassroots networks working together to promote the messages of solidarity, sustainability and emancipation collectively led to the success of the Chipko Movement, which today is recognised as "a transnational icon for environmental action" (Chatterjee, 2012: 1). As an ecological movement, it has inspired environmental activists, eco-feminists, indigenous groups and grassroots organisers all over the world by its simple but powerful act of embracing trees.

Tasmanian dam protests – networks

The campaign to save Tasmania's Franklin and Gordon rivers, threatened by the proposal of a huge hydro-electric development, galvanised the environmental movement in Australia and led to the formation of the first political party devoted to Green politics. For many years, the Hydro-electric Commission in Tasmania maintained great influence over the Tasmanian state government. In the early 1970s, it succeeded in having three dams built on Lake Pedder, despite concerted opposition. By 1976, the Hydro decided to dam the Franklin River in Tasmania's wild south-west and in 1978 released plans for the construction of two dams. But this time their attempt spurred a group of activists, still angered by the Lake Pedder dams, who took their opposition to the plan Australia-wide, and to the United Nations.

The Tasmanian Wilderness Society, formed in the wake of the Lake Pedder Dams fight, together with the Tasmanian Conservation Trust and the Australian Conservation Foundation, mounted an Australia-wide public interest campaign against the Hydro, using film and photography of the region. Bob Brown, a Launceston doctor and environmental activist, filmed his journey in an inflatable raft with Paul Smith, a local forester, down the threatened Franklin in 1976. The film was shown on Tasmanian television and helped to create awareness of what might be lost if the Hydro's plans succeeded. Documentary films played an important role in raising awareness among ordinary Australians who saw for the first time the untouched beauty of the region.

The community was deeply divided. Such was the power of the Hydro-electric Commission that the environmental movement was opposed by all three Tasmanian newspapers, all but one trade union, both houses of Tasmanian state parliament and the business sector. In June 1980, ten thousand people marched in Hobart as protests over the proposed dams increased. Both sides engaged in blitz advertising. Across Australia, membership of the anti-dam movement increased exponentially. The triangular 'No Dams' sticker was displayed everywhere and rallies were held across the country.

ABC radio's classical music station held a 'Concert for the Franklin' while Bob Brown and British botanist and TV presenter, David Bellamy, hosted a candle-lit dinner for five hundred guests, eradicating any further suggestion that the protesters were just a bunch of hippies. A song, 'Let The Franklin Flow', became the movement's campaign anthem. The blockade of the dam site resulted in 1217 arrests, nearly 500 of whom were jailed. In 1981, Aboriginal caves were discovered in the area containing stone tools and stencils, which were up to twenty four thousand years old. In 1983 the new federal government, under Labour leader Bob Hawke, passed legislation and regulations overriding the Tasmanian government, but Tasmania ignored the new law. The battle went to the High Court of Australia in May 1983. By four votes to three the High court found in favour of the federal government, that it had the power to override the states.

The movement gave birth to a political party, the Australian Greens, with Bob Brown its leader elected to the Tasmanian parliament. Brown and his friends had not only won, despite having endured incarceration and assault. They had changed environmental activism forever, by appealing not only to the young student class, but to all thinking people, regardless of class or political persuasion. The methods used included all forms of media then available: TV, cinema, radio, newspapers, flyers, arresting logos, as well as rallies, music and celebrities.

The Wilderness Society, by now a national movement, became the model for wilderness groups all over the world, and a new understanding of the importance of environmental responsibility had begun. At least as importantly, Brown's co-founding of the new political party, the Tasmanian Greens, resulted in the expansion of Green politics nationally, then internationally. Brown is no longer a politician, but remains active in the environmental movement.

Standing Rock Sioux – diversity

Since 2016 indigenous groups have been protesting in North Dakota against the North Dakota Access Pipeline (DAPL), a $3.7 billion project being delivered by the Dallas company, Energy Transfer Partners. The footprint of the pipeline project is set to cross the Missouri River, Cannon Ball River and Lake Oahe, threatening the water source and sacred land of a Native American Tribe, the Standing Rock Sioux (DiChristopher, 2017).

In response to the DAPL one of the largest environmental campaigns began, advocating for the environment and the rights of the Standing Rock Sioux. The campaign against the DAPL focused primarily on encouraging community participation. The grassroots campaign used a number of traditional and non-traditional communication channels in order to rally against the government, in a bottom-up environmental communication approach. Indigenous activists set up a 'Spiritual Camp' on the banks of Missouri River in April 2016, the first of a number of camps set up during the protests.

A media outreach strategy was employed to raise awareness of the protests taking place, and to garner support for the cause. Social media was utilised, with the hashtag: '#NoDAPL' trending on twitter. Thousands of supporters were mobilised in the largest gathering of Native Americans in more than 100 years (Wong, 2016). National and international environmental activists, Members of Congress, celebrities and veterans joined the campaign to protest against the DAPL. GoFundMe.com pages were set up, and petitions on Change. Org emerged, with one GoFundMe page raising over $1 million in five months, and one petition collecting over 556,000 signatures (Lee, 2016). Further, the 'Rezpect our water!' campaign was founded, with the youth of the Standing Rock Sioux organising a cross-country run to deliver a petition to the US Army Corps of Engineers (Rezpect Our Water, 2017).

Standing Rock protests, through platforms such as Facebook and Twitter, successfully incorporated social media as a new form of participatory strategy. During the Standing Rock protests, over one million global citizens used the location-based 'Check-In' option offered by Facebook to aid in global acknowledgement of the cultural and environmental concern in the region (Levin and Woolf, 2016). While the effectiveness of social media is still being evaluated, especially in relation to political pressure and policy reform, its impact in this specific case legitimises its ability to transform a local issue into one that is recognised on a global scale. In the case of the Dakota Access Pipeline concerns, social media played a major role as a participatory media outlet for the local communities.

The DAPL campaign taught many lessons which will be useful for future environmental movements. Frontline organisers credit networking amongst diverse groups as one of the main strengths of the protest movement. These networks were reached through social network sites which elicited the support of professionals including lawyers, teachers, academics and actors alongside the various well-established groups such as Black Lives Matter and the LGBTQI community.

* * *

Environmental challenges in the Pacific

Collectively the Pacific Islands are the most vulnerable in the world. They suffer floods and droughts controlled by the El Niño Southern Oscillation (ENSO) phenomenon. Drought and flood risks change during La Niña (normal cool years) and El Niño (warmer years) affecting different areas of the Pacific in different ways. For example, La Niña brings flood risk in Western Pacific and drought in Central Pacific, and during El Niño it is reversed with flood risk in Central Pacific and drought in Western Pacific.

Hydro-meteorological hazards most common in the Pacific are ocean acidification and temperature increase, which means more warm days and warm nights and fewer cold days and cold nights. Climate change affects the frequency and intensity of natural hazards and will, then, impact on people's ability to respond effectively. Two recent super cyclones have devastated homes, infrastructure and the economies of Fiji and Vanuatu affecting the livelihood of almost half of their populations. The region faces other natural hazards not linked to climate change – volcanoes, tsunamis and earthquakes – due to its location in the Pacific Ring of Fire (McNamara et al., 2012).

Despite the fact that Pacific island countries (PICs) emit less than 0.01% of total global greenhouse gas responsible for climate change, the region is among the most vulnerable in the world to its negative impacts (Oxfam, 2015; WWF Pacific, 2015). More than 50% of the region's population lives within 1.5 km of the shore, and many of these countries are barely a few metres above sea level, putting them at extreme risk from sea level rise. Changes are already being recorded in the region's climate and ocean. Parts of the western Pacific have experienced the fastest rate of sea level rise in the world at 10mm per year (SPREP, 2013). Sea level is projected to rise by more than

1 metre by 2100 (Nunn, 2013), and affect communities living along the coasts as well as the coastal ecosystems and natural resources that people depend on for their survival. Permanent cross-border resettlement is one of the projected impacts of climate change for some small island states, among them Tuvalu, Kiribati and the Marshall Islands, where their populations are now coming to terms with the prospect of becoming climate refugees (McNamara and Gibson, 2009; Yamamoto and Esteban, 2014). Kiribati has already made provision for the future by purchasing land in Fiji.

Rising sea levels are just one of the many effects of climate change in the Pacific Islands. Associated impacts of climate change, both currently experienced and projected, threaten communities' livelihoods and long-term viability. These include an increase in the frequency and intensity of cyclones, coral bleaching, coastal erosion, changing patterns of pests and diseases, saltwater intrusion, storm surges and flooding, increased temperatures (affecting taro production on lowlands and increasing risk of fire) and drought. Irregular rainfalls, resulting in floods and droughts, will have a devastating impact on agriculture production, forcing reliance on imported foods and threatening the livelihood of local farmers (SPREP, 2013).

Pacific islanders face numerous home-grown ecological challenges. These include poor urban planning, waste disposal, energy inefficiency and overuse of natural resources. Social and economic factors facing urban populations in Pacific island countries include urban drift, leading to overpopulation in cities and towns, insecure housing, high cost of living leading to poor nutrition and health, which add to the burden of recovery from hazards. Some nations are still being impacted by ecological disasters inherited from colonial rule.

The colonial legacy and nuclear testing

In the Pacific, like other developing regions of the world, the environmental movement was framed within the larger emancipatory politics for self-determination and indigenous rights to land and natural resources. Between 1946 and 1996 Oceania was the site for nuclear testing by the United States, Britain and France. A total of 315 atmospheric and underground tests were carried out in the colonial territories of the Marshall Islands, Gilbert and Ellis Islands (now Kiribati and Tuvalu) and the Mururoa Atoll. Alister Barry's documentary 'A Nuclear Free Pacific' (1988) makes the point:

> The view of the South Pacific is that of a region of peace. The reality is that the South Pacific has played the role of midwife to the

> nuclear arms race. The irony is that it was a role imposed on the region by its allies.

The colonial powers had little regard for the wellbeing of Pacific island inhabitants who were under their jurisdiction. Instead the Oceanic region, with 96% sea surface and thousands of uninhabited islands, was seen as a 'safe zone' because of its distance from the metropolitan centres of the world. Smith (1997: 1) explains that during the post-war period the micro-states of the Pacific were generally dismissed as 'political backwaters' . . . and Pacific islanders portrayed as passive actors – as opposed to recognising their abilities to influence events.

Within the region there was rising concern about nuclear fallout and its health implications on Pacific island communities and marine life that was their main source of food. In 1975 the first conference of the Nuclear Free and Independent Pacific (NFIP) in Suva, Fiji brought together church leaders, academics, students, emerging political leaders, environmental activists and indigenous community groups (see Robie, 2014). As a young student in Papua New Guinea at the time, Rex Rumakiek campaigned to raise awareness about the West Papua Independence movement. He remembers:

> It was an exciting time. People really felt connected to the issue. It was inspiring to go to the meeting. There were fights, there were cries because people felt like they were discovering something new, something that could solve the problems. They found their voice and their networks through this movement.
>
> (R Rumakiek, 2016, personal communication, 4 November)

Rumakiek, who later became one of the leading campaigners for the West Papua independence movement, recalls the meeting as a historically important event that marked the emergence of a collective Pacific consciousness.

> The people realised that these issues are connected – the nuclear issue, the danger of nuclear and colonisation, these issues are connected. The fact that these devices are being tested on indigenous land is because people have no freedom to make decisions, they are still under a colony so they need to be liberated, they need their freedom so they can stop these devices to be tested on their ground.
>
> (R Rumakiek, 2016, personal communication, 4 November)

The meetings marked a watershed in Pacific political consciousness and gave rise to the environmental movement which sought an end to nuclear testing in the region. As political awareness grew within the region throughout the '70s and '80s, Pacific islanders asserted their own political agenda, channelling their efforts through the Nuclear Free and Independent Pacific movement. NFIP was a loose collective of civil society representatives who formulated The People's Charter for Nuclear Free and Independent Pacific in 1983. It states in its Preamble: "We, the people of the Pacific will assert ourselves and wrest control over the destiny of our nations and our environment from foreign powers, including the Trans-National Corporations" (NFIP, 1983). A book titled *A Call to a New Exodus: An Anti-Nuclear primer for Pacific People* was published by the Pacific Conference of Churches to educate and raise awareness about the dangers of nuclear testing. It urged concerted grassroots action. "It is possible for ordinary people to make a difference. It is **more** possible if individuals unite themselves in groups" (Siwatibau and Williams, 1982: 86, bold in original).

The French government met rising opposition to its activities in the Pacific with military-style response when in 1985 it bombed the Greenpeace anti-nuclear flagship, the *Rainbow Warrior*, which was moored in Auckland Harbour, New Zealand. David Robie, a Pacific-based journalist and media educator, who has covered nuclear protests in the region and written two books about the *Rainbow Warrior* bombing, observes:

> The *Rainbow Warrior* affair in 1985, involving state terrorism by a friendly nation, became iconic in New Zealand history because it highlighted New Zealand opposition to nuclear testing in the Pacific on a global stage. It also put the spotlight on solidarity with smaller nations in the Pacific that had courageously stood up to big power nuclear politics.
>
> (Robie, 2014: 259)

Despite international condemnation French nuclear testing continued in the Pacific until 1996. France maintains its jurisdiction over French Polynesia, New Caledonia and Wallis and Futuna known as 'Overseas Territories of France'. On the Marshall Islands today, a nuclear waste dump site left by United States is the new environmental time bomb. As sea levels rise nuclear waste from the massive radioactive dump site is leaking into the surrounding sea water. In a cruel twist of fate, the remnant of past colonial excesses, entombed in the sand, has resurfaced to compound the current environmental crisis, once again threatening the lives and livelihood of a new generation of the Marshallese population.

Mining and ecological disasters

The Pacific region has experienced other conflicts between the local population and foreign interests arising from environmental mismanagement of land and water in the extraction of mineral resources. Papua New Guinea endured ecological destruction on a huge scale as a direct result of bad mining practices in the Bougainville and Ok Tedi mines (Kirsch, 2007, 2014). One of the worst human-caused ecological disasters occurred at the Ok Tedi copper-gold mine in the western province of Papua New Guinea operated by the Australian mining giant then called Broken Hill Proprietary (BHP, now BHP Billiton). An average of 50 million tonnes of tailings and mine waste per year were dumped in the Ok Tedi and Fly Rivers, causing large-scale damage to the surrounding forests and wildlife, poisoning the river system and village gardens, and leading to the displacement of villagers from their traditional lands. A class action by Papua New Guinean landowners forced BHP to commission a report into its operations at Ok Tedi mines. In 1996 the company reached an out of court settlement with the landowners for an estimated US $500 million. With BHP's withdrawal from the mine its commitment to clean up the polluted areas remains to be met (Troubled Waters, 2012).

In Bougainville, conflict ensued between landowners and mine operators as a result of pollution from mine tailings to the surrounding environment and its impact on the islanders' traditional way of life. Panguna Mine, one of the world's biggest open cut copper mines, was owned and operated by a subsidiary of the mining giant Rio Tinto. While the mine generated billions of dollars of revenue for the PNG government and huge profits for the mine operators, Bougainville landowners received little compensation for their land and rivers, which were being poisoned as reported in the *Sydney Morning Herald*: "millions of tonnes of acid-laced mine tailings killed the Jaba and Kawerong rivers. The rivers had been a source of water and food for thousands, but large sections now resemble a moonscape, forcing people to leave their homes" (Flitton, 2016). With their demands unmet, a rebel group made up of young Bougainville landowners formed the Bougainville Revolutionary Army (BRA). They fought a bloody civil war which drove out both the PNG government forces and Rio Tinto Zinc (RTZ) leading to the closure of the mine (see Robie, 2014; Phillips, 2015). The decade-long conflict, which began in 1988 and cost 20,000 lives, became known as the first successful eco-revolution, and its story is told in the award-winning documentary *The Coconut Revolution* (2000). The Autonomous Bougainville Government continues to seek environmental justice from mining polluters to clean up large areas of the land that remain uninhabitable (Flitton, 2016).

In West Papua, similar concerns are recorded in the operation of the Freeport copper mine, where 200,000 tonnes of tailings per day (over 80 million tonnes per year) are dumped in the river systems, causing floods, deforestation and grave impacts on community health and livelihoods (Troubled Waters, 2012). In the case of West Papua, which is under Indonesian rule, the mining operations not only imperil the environment, but also the indigenous population who are considered to be at risk of becoming "an anthropological museum exhibit of a bygone culture" (Schulman, 2016).

The West Papua resistance to Indonesian rule has claimed thousands of lives, however the international community including Australia and the United States continue to recognise Indonesian sovereignty in the region. Many small island nations rely exclusively on their river systems for fresh water. Protection of rivers and streams from long-term pollution is more critical than ever before, given the increased presence and severity of droughts due to climate change.

Nauru and Banaba Island in Kiribati are other examples of Pacific nations that have borne the consequences of decisions made by colonial powers with vested commercial interest, which disregard the rights of indigenous people under their protection. Both these countries have suffered irreversible environmental damage from intensive phosphate mining with loss of 80% of island soil from strip mining. Much of the mined phosphate has become part of Australian farmland as agricultural fertiliser. After having experienced the ups and downs of its mining fortunes, Nauru now earns its income by hosting the much maligned offshore refugee detention centre for the Australian government. In the aftermath of World War II, colonial authorities moved the entire population of Banaba to Rabi Island in Fiji because the island home was rendered uninhabitable. As Jane McAdam observes,

> the colonial administration noted that while it would be "repugnant" to "evict a native tribe" simply "to afford wider opportunity of gain to a rich commercial corporation", in this case the greater good of the British Empire was at stake.
>
> (McAdam, 2013)

McAdam draws a comparison between the forced relocation of the Banaba population with the current scenario of moving an entire Pacific island population as a consequence of rising sea levels due to climate change. She cautions that in case of future relocation "the rights of those affected (both in the sending and receiving countries) must be protected, and the legal status and organisational structures of the relocated group in the new country must be planned meticulously" (McAdam, 2013).

Networks and Pacific regionalism

Pacific scholars have searched for their unique Pacific identity borne out of Pacific indigenous histories in order to counteract the neo-colonial imagination. The imperialists envisioned the Pacific island countries as small, remote, disconnected and dependent. These nation states were in need of salvation by their economically powerful neighbours, especially "Australia which saw itself as the natural leader of the post-colonial South Pacific" (Fry, 1996: 2).

The interconnectedness of islands has been an important framework of analysis in Pacific island studies. In 1994, eminent Pacific scholar Epeli Hau'ofa argued that the Oceanic region should be imagined as 'Our sea of islands' with the Pacific Ocean as their common home (Hau'ofa, 1994). In reimagining the region Hau'ofa recalled the pre-colonial links of Pacific cultures, their mastery of the ocean, and the myths and legends that guided them, as he observed:

> But if we look at the myths, legends, and oral traditions, and cosmologies of the peoples of Oceania, it becomes evident that they did not conceive of their world in such microscopic proportions. Their universe comprised not only land surfaces, but the surrounding ocean as far as they could traverse and exploit it, the underworld with its fire-controlling and earth-shaking denizens, and the heavens above with their hierarchies of powerful god and named stars and constellations that people could count on to guide the ways across the seas.
>
> (1994: 154)

Hau'ofa's call to reframe the Pacific included not only dry land surfaces but the oceans that surrounded them because, "There is a world of difference between viewing the Pacific as 'islands in a far sea' and as 'a sea of islands'" (Hau'ofa, 1994). A new generation of Pacific leaders asserts this revised regional identity of an 'Oceanic continent' echoing Hau'ofa's imaginary identity, which favours the interconnections of the Pacific island states. New Pacific regionalism has emerged despite pressure from the metropolitan neighbours. Maureen Penjueli, co-ordinator of the Pacific Network on Globalisation, observes:

> Continued attempts by the Forum Secretariat to frame the region consistently as small, vulnerable, fragile, aid-dependent is no longer a working narrative for our people let alone our countries. The wind of self-determination has set our countries on a journey to reduce our dependency on foreign aid, to be more self-sufficient, to negotiate and demand a fair share of the wealth of our resources, to redefine security and find newer partners (rightly or wrongly).
>
> (Penjueli, 2017)

Instead of isolation, the surrounding sea presents new economic opportunities for the ocean states, especially the benefits generated under the exclusive economic zones (EEZ) (See Fig 4.1). Some of the smallest islands today command some of the largest ocean resources under the terms prescribed by the United Nations Convention on the Law of the Sea. This includes the 370 km boundary over which a state has special rights regarding the exploration and use of marine resources, including energy production from water and wind (Rona, 2003). Kiribati today controls one of the largest EEZ in the world, embracing fisheries stock and future prospects of mining the ocean floor. Ironically this brings with it new threats of domination and exploitation by more powerful states which are already manoeuvring for favourable positions through political and soft power diplomacy. It also raises the ugly spectre of a new frontier for environmental disasters with growing interest in seabed mining and exploration.

In the 21st century Pacific island nations are strengthening their networks across a wide spectrum of socio-cultural, economic, political and scientific endeavours (Harris, 2008, 2014). This solidarity has been most pronounced in the area of climate change negotiations. In 2015 Leaders of Pacific Islands Development Forum signed the Suva Declaration on Climate Change which called for an agreement 'to stabilise global average temperature increase to well below 1.5°C above pre-industrial levels' (Pacific Islands Development Forum, 2015: 1). The Declaration reflects the concerns of all Pacific islands nations by noting that "climate change poses irreversible loss and damage to our people, societies, livelihoods and natural environments; creating existential threats to our very survival and other violations of human rights to entire Pacific Small Island Developing States" (Pacific Islands Development Forum, 2015: 1).

The main positions put forward in the Suva Declaration were reflected in the final Paris Climate Change Agreement of the 21st session of the Conference of Parties (COP21) in Paris in 2015. The Suva declaration for the first-time enabled Pacific island nations to put a cohesive Pacific voice in climate change negotiations. The strong-arm tactics of its more powerful neighbour, Australia, which had tried to water down the Pacific leaders' statement on climate change, failed to work. Fiji's presidency at COP23 in Bonn, Germany in 2017 is further proof of a strengthening Pacific diplomacy on the world stage.

Bottom-up networks

At another level, regional civil society organisations (CSO) are proactive in environmental advocacy. The Pacific Island Climate Action Network (PICAN) advocates for climate justice at regional and international forums. As part of a three-tiered network, PICAN is made up of smaller country-based networks – e.g. VCAN in Vanuatu, KiriCAN in Kiribati and TCAN in Tuvalu – and is linked to the worldwide Climate Action Network (CAN) which has over 900 non-government organisations "working to promote government and individual action to limit human-induced climate change to ecologically sustainable levels" (Climate Action Network, 2017; Pacific Island Climate Action Network, 2017). Pacific Climate Warriors from 350.org Pacific enlist youth to take global action on climate change. In 2014, Pacific Climate Warriors paddled out in traditional canoes to blockade the world's largest coal port in Newcastle, Australia. Through their banner 'we are not drowning, we are fighting', Pacific Climate Warriors reframe the narrative to show that Pacific islanders are not passive victims but proactive civil society actors demanding climate justice.

The 'bottom-up regionalism' which once forged the nuclear-free movement is now seeing a resurgence around climate change advocacy led by civil society actors (Titifanue et al., 2017: 142). CSOs organised protests at the annual climate change discussions during the COP23 in Bonn and the crucial Paris Agreement at COP21. Social media platforms allow groups to "share information, communicate and organise" using multimedia content, event invitations to protest sites, and use of hashtags using local and global networks. Titifanue et al. observe that after the anti-nuclear movements of the 1980s

> regional solidarity seemed to take a hiatus at the grassroots level and became simply something that decision makers and academics concerned themselves with. In recent times however, through

> digital interconnectivity, there has been resurgence in grassroots interest on key issues that have captivated the public imagination.
> (2017: 142)

Like the nuclear testing of 50 years ago, Pacific peoples are once again angry at the injustice of being at the receiving end of environmental impacts which are not of their making. A Pacific regionalism at both top and bottom levels of society has emerged to organise and protest against the excesses of the developed world.

Conclusion

Environmental movements seek to protect Earth's ecological system from human activities which cause long-term detrimental impact on land, water, air and the wellbeing of all living things. Vertical communication is effective in disseminating information about environmental concerns and in organising strategic action. Horizontal communication encourages greater dialogue, allowing participants to listen to each other, share opinions and debate issues with the aim of finding solutions.

Pacific islanders are revitalizing old alliances using new communications technologies in the face of numerous environmental challenges. Participatory communication networks enable communities to share knowledge and create awareness about a range of issues from their own experiences. The networks have the ability to change the conversation from alarmist notions of climate change to a more practical discussion of impacts that are both real and immediate, such as access to energy, food and water security, sanitation, jobs and livelihoods (Faris et al., 2014). Communication between participants empowers endogenous networks and uncovers unique solutions arising from the particular context of the community.

References

A Nuclear Free Pacific 1988. Documentary. Produced by Alister Barry and Philip Shingler; Written and Directed by Lesley Stevens. Retrieved from https://www.nzonscreen.com/title/nuclear-free-pacific-1988/credits

Anderson, A. 2015. Reflections on environmental communication and the challenges of a new research agenda. *Environmental Communication*, 9(3): 379–383. doi:10.1080/17524032.2015.1044063

Bandyopadhyay, J. 1999. Chipko movement: Of floated myths and flouted realities. *Economic and Political Weekly*, 34(15): 880–882.

Brulle, R. J. 2010. From environmental campaigns to advancing the public dialog: Environmental communication for civic engagement. *Environmental Communication*, 4(1): 82–98. doi:10.1080/17524030903522397
Carson, R. 1962. *Silent Spring*. Boston: Houghton Mifflin.
Castells, M. 2009. *Communication Power*. Oxford: Oxford University Press.
Chabot, S. & Vinthagen, S. 2007. Rethinking nonviolent action and contentious politics: Political cultures of nonviolent opposition in the Indian independence movement and Brazil's landless workers movement. *In:* Coy, P. G. (ed.) *Research in Social Movements, Conflicts and Change*. Bingley, West Yorkshire: Emerald Group Publishing Limited, 91–121.
Chatterjee, D. 2012. *Leading Consciously*. Abingdon: Routledge.
CIVICUS. 2016. *State of Civil Society Report 2016* [Online]. Johannesburg: CIVICUS World Alliance for Citizen Participation. Retrieved from www.civicus.org/images/documents/SOCS2016/summaries/SoCS-full-review.pdf
Climate Action Network. 2017. *About CAN* [Online]. Climate Action Network International. Retrieved 17 March 2017 from www.climatenetwork.org/about/about-can
DiChristopher, T. 2017. Trump ignores question about Standing Rock Sioux after signing Dakota Access order. *CNBC*, 24 January. Retrieved from www.cnbc.com/2017/01/24/trump-ignores-standing-rock-sioux-question-after-dakota-access-order.html
Downing, J. D. 2001. *Radical Media: Rebellious Communication and Social Movements*. Thousand Oaks, CA: Sage Publications.
Dutta, M. J. 2012. *Voices of Resistance: Communication and Social Change*. West Lafayette, IN: Purdue University Press e-Books.
Faris, S., Lipscombe, S., Whitehead, S. & Wilson, D. 2014. *From the Ground Up: Changing the Conversation about Climate Change* [Online]. London, UK: BBC Media Action. Retrieved from http://downloads.bbc.co.uk/mediaaction/policybriefing/bbc_media_action_from_the_ground_up_climate_change.pdf
Flitton, D. 2016. Rio Tinto's billion-dollar mess: "unprincipled, shameful and evil". *Sydney Morning Herald*, 21 August. Retrieved from www.smh.com.au/world/billiondollar-mess-a-major-disaster-the-people-do-not-deserve-to-have-20160817-gquzli.html
Fry, G. 1996. *Framing the Islands: Knowledge and Power in Changing Australian Images of "the South Pacific"*. Canberra: National Library of Australia.
Habermas, J. 1989. *The Structural Transformation of the Public Sphere: An Inquiry into a Category of Bourgeois Society*, Cambridge, MA: The MIT Press.
Harris, U. S. 2008. Bridging the divide with participatory video. *Fijian Studies: A Journal of Contemporary Fiji*, 6(1/2): 146–165.
Harris, U. S. 2014. Communicating climate change in the Pacific using a bottom-up approach. *Pacific Journalism Review*, 20(2): 77–96. doi:10.24135/pjr.v20i2.167
Hau'ofa, E. 1994. Our sea of Islands. *The Contemporary Pacific*, 6(1): 148–161.

Ibhawoh, B. & Dibua, J. 2003. Deconstructing Ujamaa: The legacy of Julius Nyerere in the quest for social and economic development in Africa. *African Journal of Political Science*, 8(1): 59–83.

Jain, S. 1984. Women and people's ecological movement: A case study of women's role in the Chipko movement in Uttar Pradesh. *Economic and Political Weekly*, 19(41): 1788–1794.

Johnson, S. 2012. *UNEP the First 40 Years: A Narrative*. Nairobi: UNON/Publishing Section Services.

Kirsch, S. 2007. Indigenous movements and the risks of counterglobalization: Tracking the campaign against Papua New Guinea's Ok Tedi Mine. *American Ethnologist*, 34(2): 303–321. doi:10.1525/ae.2007.34.2.303

Kirsch, S. 2014. *Mining Capitalism: The Relationship between Corporations and Their Critics*. Berkeley, CA: University of California Press.

Kretser, H. D. 2016. NSW anti-protest laws are part of a corrosive national trend. *The Sydney Morning Herald*, 22 March 2016. Retrieved from www.smh.com.au/opinion/nsw-antiprotest-laws-are-part-of-a-corrosive-national-trend-20160322-gno10h.html

Krinsky, J. & Crossley, N. 2014. Social movements and social networks: Introduction. *Social Movement Studies*, 13(1): 1–21. doi:10.1080/14742837.2013.862787

Lee, A. J. B. 2016. *Stop the Dakota Access Pipeline* [Online]. Change.Org. Retrieved 15 April 2017 from www.change.org/p/jo-ellen-darcy-stop-the-dakota-access-pipeline

Levin, S. & Woolf, N. 2016. A million people "Check In" at standing rock on Facebook to support Dakota pipeline protesters. *The Guardian*, 1 November 2016. Retrieved from www.theguardian.com/us-news/2016/oct/31/north-dakota-access-pipeline-protest-mass-facebook-check-in

McAdam, J. 2013. Caught between Homelands. *Inside Story*. Retrieved from http://insidestory.org.au/caught-between-homelands/

McNamara, K. E. & Gibson, C. 2009. We do not want to leave our land: Pacific ambassadors at the United Nations resist the category of climate refugees. *Geoforum*, 40(3): 475–483. doi:10.1016/j.geoforum.2009.03.006

McNamara, K. E., Hemstock, S. L. & Holland, E. A. 2012. *PACE-SD Guidebook: Participatory Vulnerability and Adaptation Assessment* [Online]. Suva, Fiji: Pacific Centre for Environment and Sustainable Development, The University of the South Pacific. Retrieved from http://eugcca.usp.ac.fj/Portals/0/PACE-GB.pdf

Milan, S. 2013. *Social Movements and Their Technologies: Wiring Social Change*. New York: Palgrave Macmillan.

Nelson Institute for Environmental Studies. 2016. *Discover the Earth Day History* [Online]. Madison, WI: Nelson Institute for Environmental Studies. Retrieved 17 January 2016 from www.nelsonearthday.net/earth-day/index.php

NFIP. 1983. *People's Charter for a Nuclear Free and Independent Pacific* [Online]. Nuclear Free and Independent Pacific, Moana Nui Action

Alliance. Retrieved from http://imipono.org/2012/09/20/peoples-charter-for-a-nuclear-free-and-independent-pacific/

Nunn, P. D. 2013. The end of the Pacific? Effects of sea level rise on Pacific Island livelihoods. *Singapore Journal of Tropical Geography*, 34(2): 143–171. 10.1111/sjtg.12021

Oxfam. 2015. *A Question of Survival: Why Australia and New Zealand Must Heed the Pacific's Calls for Stronger Action on Climate Change* [Online]. Port Moresby: Oxfam. Retrieved from www.oxfam.org.au/wp-content/uploads/2015/09/2015-86-pacific-islands-forum-breifing-a-question-of-survival_fa2.pdf

Pacific Island Climate Action Network. 2017. *Vision* [Online]. Pacific Island Climate Action Network. Retrieved 17 March 2017 from https://pacificclimateactionnetwork.wordpress.com/about-us-2/vission/

Pacific Islands Development Forum. 2015. *Suva Declaration on Climate Change* [Online]. Suva, Fiji: Pacific Islands Development Forum. Retrieved from http://pacificidf.org/wp-content/uploads/2013/06/PACIFIC-ISLAND-DEVELOPMENT-FORUM-SUVA-DECLARATION-ON-CLIMATE-CHANGE.v2.pdf

Penjueli, M. 2017. *State of Pacific Regionalism: Is the Current Regional Architecture Fit for Purpose for the Complexities of the 21st Century?* [Online]. Retrieved from 12 May 2017 www.pina.com.fj/index.php?p=pacnews&m=read&o=139927374559af6542482380179b6d

Phillips, K. 2015. *Bougainville at a Crossroads: Independence and the Mine* [Online]. ABC Radio National. Retrieved from 16 March 2017 www.abc.net.au/radionational/programs/rearvision/bougainville-at-a-crossroads/6514544

Rezpect Our Water. 2017. *Rezpect Our Water: Stop the Dakota Access Pipeline* [Online]. Rezpect Our Water. Retrieved 16 April 2017 from http://rezpectourwater.com/

Robie, D. 2014. *Don't Spoil My Beautiful Face: Media, Mayhem & Human Rights in the Pacific*. Auckland: Little Island Press.

Rona, P. A. 2003. Resources of the sea floor. *Science*, 299(5607): 673–674. doi:10.1126/science.1080679

Schulman, S. 2016. The $100bn gold mine and the West Papuans who say they are counting the cost. *The Guardian*. Retrieved from www.theguardian.com/global-development/2016/nov/02/100-bn-dollar-gold-mine-west-papuans-say-they-are-counting-the-cost-indonesia

Siwatibau, S. & Williams, D. 1982. *A Call to a New Exodus: An Anti-Nuclear Primer for Pacific People*. Suva, Fiji: Pacific Council of Churches.

Smith, R. H. 1997. *The Nuclear Free and Independent Pacific Movement after Muroroa*. London: Tauris Academic Studies.

SPREP. 2013. *Climate Change* [Online]. Apia, Samoa: Secretariat of the Pacific Regional Environment Programme. Retrieved from www.forumsec.org/resources/uploads/attachments/documents/3-%20Climate%20Change%20SDWG%20Brief%20(8Mar13).docx

The Coconut Revolution 2000. Documentary. Directed by Dom Rotheroe; Produced by Darren Bender and Mike Chamberlain. Stampede Films.

The Great Smog of 1952. 2015. *The Great Smog of 1952* [Online]. Met Office. Retrieved 17 March 2017 from www.metoffice.gov.uk/learning/learn-about-the-weather/weather-phenomena/case-studies/great-smog

The History of Earth Day. 2016. *The History of Earth Day* [Online]. Earth Day Network. Retrieved 17 Jaunary 2016 from www.earthday.org/about/the-history-of-earth-day/

Titifanue, J., Kant, R., Finau, G. & Tarai, J. 2017. Climate change advocacy in the Pacific: The role of information and communication technologies. *Pacific Journalism Review*, 23(1): 133–149.

Troubled Waters. 2012. *Troubled Waters: How Mine Waste Dumping Is Poisoning our Oceans, Rivers, and Lake* [Online]. Washington, DC: Earthworks and MiningWatch Canada. Retrieved 17 March 2017 from www.earthworksaction.org/files/publications/Troubled-Waters_FINAL.pdf

WCED. 1987. *Our Common Future (Brundtland Report)*. Oxford: United Nations World Commission on Environment and Development, Oxford University Press.

Wong, J. C. 2016. Dakota access pipeline: US denies key permit, a win for Standing Rock protesters. *The Guardian*, 5 December. Retrieved from www.theguardian.com/us-news/2016/dec/04/dakota-access-pipeline-permit-denied-standing-rock

WWF Pacific. 2015. *Time Is Running Out for Low-Lying Islands in the South Pacific* [Online]. Honiara, Solomon Islands. Retrieved 20 November 2017 from www.wwfpacific.org/what_we_do/climatechange/

Yamamoto, L. & Esteban, M. 2014. *Atoll Island States and International Law: Climate Change Displacement and Sovereignty*. Heidelberg, Germany: Springer-Verlag.

6 Community informatics and the power of participation

Reflection

In the 1970s my mother used to traverse many miles of dusty roads to encourage rural women to participate in self-improvement and community activities. Although she had a telephone, her compatriots did not. So, her tools were the local bus, her feet and a deep commitment to personal empowerment. Decades later when I conducted the participatory media training in Fiji, only a few rural women owned mobile phones. One afternoon as I sat chatting with a group of women, I was startled by the sound of a cow mooing next to me. I searched for the elusive cow only to find that it was the ringtone of a mobile phone, greatly amusing the women. People always find ways to adapt technology to their local context. After the participatory video training, the women reflected on the capacity-building aspect of the training and their aspirations to become part of technological change: One confided, "This day I'll never, never forget in my life". Another described it as a "privilege". Most of all the knowledge they had gained had also quietly made them confident about the Information Age, as one of the participants said to me, "this is the era of technology and we've been invited to become part of it".

Today, Pacific islanders have easier access to mobile technologies than to landlines. Facebook and twitter are important everyday tools of communication. Costs are still high, but many in the community would rather cut back on their meals than give up access to mobile data.

Introduction

Information and communication technology (ICT) provides a vital lifeline for vulnerable communities during major disasters. People caught in the midst of cyclones, forest fires and floods rely on social media such as Facebook and Twitter, and the use of mobile phone data, as the first point of contact with the outside world. Access to ICT is a necessity in disaster risk reduction (DRR) and an important resource in climate change adaptation in both urban and rural contexts.

This chapter discusses the use of ICTs for climate change communication in the Pacific. It demonstrates how strategic integration of participatory media at community level encourages greater citizen engagement by enabling them to record, share and discuss environmental issues in their own environment. It also provides some innovative examples of ICT responses to disaster risk reduction and other environmental challenges by governments and regional organisations.

ICTs are important resources in solving complex environmental problems such as climate change (Yap, 2011; Cave, 2012). ICTs are defined as "an expanding assembly of technologies that can be used to collect, store and share information between people using multiple devices and multiple media" (Chapman and Slaymaker, 2002: 1). ICTs offer a powerful way of inviting wider representation of people's voices and views, which are necessary for democratic participation. Governments in developing countries see the adoption of new technologies such as the Internet as a panacea for their nations' economic ills and as essential in creating prosperous information societies. But to harness the full potential of ICT, it is imperative that communities, including the most marginalised, have meaningful engagement with the relevant technology to be able to express effectively their needs and aspirations. Unequal access to ICT remains the greatest challenge. Despite recent gains, the least developed countries still lag behind in mobile broadband subscriptions, with women having least access to ICTs (ITU, 2017).

The communication context

Convergence of conventional technologies such as telephone, radio and television with newer forms such as the Internet, wireless and satellite technologies, coupled with computers and mobile devices, have delivered easier access to information and participation in content creation. Research exploring the links between ICT, climate change and development has identified four main areas of ICT use – mitigation, monitoring, adaptation and strategy (Ospina and Heeks, 2010). The UN's Food & Agriculture Organisation identified three main functions of ICT: "(1) to record data and information, (2) to transform the data and information into knowledge which can be shared and, eventually (3) communicate the data, information and knowledge" (Sala, 2010: 3). Appropriate ICTs for specific situations and communities play a primary role in efforts to "develop, plan, implement and manage adaptation strategies", and to improve communication within and between communities (Sala, 2010: 6; Yap, 2011). Information-based technologies are most effectively used in mapping long-term climate change and monitoring short-term climate variability, whereas communication-based technologies assist in information and knowledge transfer (Sala, 2010).

In adaptation processes ICTs are most effective in networking and knowledge exchange of good practices, inclusion of vulnerable communities in decision-making for best policy outcomes, and integration of information and knowledge that have local relevance in reducing risk and vulnerability (Ospina and Heeks, 2010; Heeks and Ospina, 2018). Message creation and dissemination must be strategically integrated with stakeholder inclusion using conventional media forms (participatory video and community radio) and new media (mobile and online platforms). As tools for empowerment, ICTs provide opportunities for local communities to raise awareness of their situation, demand action and accountability from their leaders, build coalitions and share knowledge using peer-to-peer networks (Kalas and Finlay, 2009).

Improving connectivity

Pacific island countries (PICs) have experienced uneven access to broadband and mobile connectivity, with some countries like Fiji achieving high speed Internet while countries such as Kiribati experiencing slow and expensive connection. Connectivity is delivered through mobile and fixed networks with connections to satellites and undersea cables. While more Pacific nations are investing in undersea cable connections, mobile networks are favoured in driving connectivity and Internet access, due to the dispersed nature of PICs, many of which are archipelagos comprised of a large number of islands. This is consistent with global trends in mobile broadband usage, which has outpaced fixed broadband subscriptions, with prices dropping substantially in least developed countries where 35% of individuals using the Internet are young people aged 15–24 (ITU, 2017). The World

Bank (2017) projects that in the next 25 years most people in PICs will have access to faster and cheaper Internet services with increasing potential for ICT in environmental management, including monitoring, grievance redress and citizenship engagement processes.

The Pacific Region Infrastructure Facility (PRIF) highlights the following changes in the Pacific (PRIF, 2015):

1. Mobile coverage across Fiji, Samoa, Solomon Islands, Tonga and Vanuatu has jumped from less than half of the population in 2005 to 93% of the population in 2014.
2. The cost of mobile calls fell by one third between 2005 and 2014.
3. The percentage of mobile phones in Pacific households rose from 49% in 2007 to 93% in 2014.
4. International Internet bandwidth jumped over 1500% between 2007 and 2014, rising from less than 100 Megabits per second to over 1 Gigabit per second (excluding Fiji which enjoyed high bandwidth from 2000 when it was connected to a submarine cable).

The Southern Cross Cable Network links the West Coast of the United States, Hawaii, Fiji, Australia and New Zealand using high-capacity fibre-optic cable, with new connections to Tonga, Vanuatu and Samoa. This has resulted in a phenomenal rise in international Internet bandwidth. Most island nations are connected by a single submarine cable which makes them vulnerable in the case of cable damage. Some mobile network operators encourage the adoption of hybrid solutions, including both satellite and fibre-optic cable in the Pacific, thus reducing the cost of connection to many sparsely populated islands (UN-OHRLLS, 2017).

Access to improved Internet and social media brings both positive and negative impacts. The number of Facebook users in the region grew over 250% between 2011 and 2014, with 80% accessing the social media site via mobile phones (Minges and Stork, 2015). In 2015 the Government of Nauru banned access to Facebook, blaming the social media site for creating instability in the nation, citing cyber bullying and offensive posts. The president of Nauru, Baron Waqa, said the action was taken to protect the small, tightly knit community of 10,000 people from unregulated use of the social networking site, which had created tensions among friends and families (ABC News, 2015).

Digital technologies encourage greater participation in political dialogue. People in both urban and rural communities are now able to participate in debates from which they previously felt excluded. Facebook discussion groups, such as Yumi TokTok Stret in Vanuatu, and blogs provide a forum for exchange of information and opinion where all users can participate. There is a greater need for understanding how to integrate ICTs like mobile phones into media and communication plans for disaster-response technologies such as broadcast radio (Tacchi et al., 2013).

Radio remains a key media platform for communication across vast distances and audiences. Community radio is an effective medium which has enabled participation of vulnerable communities in climate change communication. A study undertaken in Malawi looked at the role of radio in promoting community participation in community-based adaptation programmes. It found that participatory radio in the form of radio listening clubs provided a viable public sphere where communities exchanged knowledge and information about climate change and decided on mitigation strategies to build community resilience (Khonje, 2011). More research is needed in the Pacific on how radio can be successfully integrated as a communication strategy in climate change education and community-based adaptation.

Socio-cultural responses to ICT

When new technology arrives in a society, it enters an already existing order of social hierarchies, power relationships and knowledge base. The adoption of the technology can enhance the position of individuals or groups in the social order and isolate those unable to co-opt it easily for their own benefit. It can widen class, gender and generational gaps. For example, young people are able to interact with new technologies more easily than their parents. Women and rural dwellers, on the other hand, benefit the least from ICT diffusion through lack of access. How does the new technology fit into the broad information ecosystem that already exists within communities? ICT use in the Pacific islands must be seen in terms of challenges faced and within the wider communication context, including the culture and belief systems of Pacific islanders.

In his discussion of technology and society, Castells (2000: 5) points out: "Technology does not determine society. Nor does society script the course of technological change . . . the final outcome depends on a complex pattern of interaction". The growing influence of technology must factor in the social and economic forces that have shaped these technologies and the powerful corporations that still have control over content and data in exchange for free access to social media platforms (Fuchs, 2009).

In developing countries, adoption is further constrained by the fact that the technologies which are introduced have been primarily developed for use in industrialised societies. As seen from recent natural disasters in the United States, the Caribbean and the Pacific region, floods and cyclones caused severe damage to primary information delivery systems due to power outages and downed mobile phone towers. Technological intervention is more effective when designed to enhance and not impede already existing forms of human communication.

Effective use of ICT should not only aid information dissemination but also invite local participation in communicative processes. It can facilitate the preservation and sharing of indigenous knowledge, provide an essential network for information and communication exchange amongst the community sector and government agencies, and enable learning and skill-building for the growing youth population. Digital narratives are a powerful way for people to capture the everyday life and concerns of a community as seen in the Papuan Voices Case Study 6.1. By using participatory media tools, village elders can share the important history and heritage of their village with the younger generation and the wider Pacific community.

Case study 6.1 Papuan Voices

By Wensislaus Fatubun
Filmmaker and founder of Papuan Voices

- *Project:* Papuan Voices is a collective movement which trains Papuan filmmakers both for empowerment and for film production. During the last six years, young West Papuans have been using video cameras to tell their own story from their own perspective. In collaboration with Engage Media, and local organisations in West Papua, we teach Papuan activists skills in video production, storytelling and distribution skills so they can tell their own stories from their own perspective to the world.
- *Context:* The conflict between West Papuans and the Indonesian Government remains largely hidden from the world.

Despite decades of hostility and violence, demands for justice have received little global attention. Papuan Voices aims to bring the life stories of West Papuans to a wider audience. These are not only the stories of conflict, they are also the stories behind the conflict: the struggles for education, the environment, equality and dignity. Importantly, the stories we tell are not only framed around West Papua's political struggle for independence. They shine light on the injustices that regularly occur behind the closed doors of this resource-rich and restive province.

- *Process:* I have developed three processes – Indigenous Mapping, Drawing technique in story, and film-making for empowerment and advocacy.
- *Indigenous Mapping* refers to the knowledge of members of a community about itself, or emic knowledge. Indigenous Mapping is based on the fact that (almost) everyone needs a visualisation of what he or she tells about himself or herself, where he or she is, what he or she sees or experiences, which he or she often describes through graffiti on the ground, or demonstrates with help. Objects around it, such as matchboxes, cigarette packs, stickers, wooden pieces, cutlery and other items. By encouraging the informant to transfer his or her natural images onto the paper, the informant's description can be 'enshrined', without having to rely on the camera. Spatial orientation of many ethno-linguistic groups (tribes) is based on an ecosystem in which members of the group are born, raised, live and buried.
- *Drawing technique in story telling* is important when one is telling his (her) story, and must be part of a collaborative work between filmmaker and communities. This 'drawing' technique can be used to complement in-depth interviewing techniques while you are filming, but be aware of some obstacles, such as informants who think 'drawing' is 'educated' work, so bring others who can draw for them, such as their 'educated' children.

From my experience, film-making is able to connect people. It is not only communication, but more of an emotional connection. People who are at different places can have an emotional connection to each other because they watch the film. Our challenge is how to produce a film about human rights that is able to connect the emotions of people and unite them to fight together. The principal instrument we have to tell the story of life is the film. I think a lot of film-making activists fail to use film as an advocacy tool because they are simply telling the issue, and not a life story.

- *Using films and film-making for empowerment and advocacy.* The process of film-making for advocacy and memory is an intersubjective experience and dialogue on story, history and culture. This is the first process when we produce a film. I visit the villages, explain what a camera is and what purpose it serves and convince the villagers of its usefulness as a tool. Once a plan for the sequence is established to everyone's satisfaction, we can film. The interest is not just in making films. Each sequence is an opportunity to talk with the villagers about human rights, the rights of indigenous peoples and their socio-economic rights. Making a film is an occasion for us to settle on a strategy with the villagers for them to claim their rights. Not only by addressing representatives of the government in meetings organised by the Justice and Peace office but also more globally using our network. We introduce the process of making a film and train the community in the early stages of building a network, we strengthen and assist the community to be able tell their stories, especially their stories about human rights.
- *Outcome:* The filmmakers see film serving a very important role for various purposes. They use film to criticise the discriminatory perspectives created by anthropologists, missionaries and government, and to provide better understanding of their society by promoting their own perspectives. They also use films as tools for documentation of historical experience, connections to nature, culture, knowledges or epistemologies, and aspirations in the future. We think that this footage should be part of ongoing advocacy process. Our videos have been used to strengthen the advocacy process in the

UN Human Rights Council. Films are also tools to consolidate our collective memory (solidarity) and for trauma healing. Papuan Voices is a cultural struggle, because we want people to see Papua through the Papuan perspective and to raise critical awareness among Papuan people. Films take Papuan voices to the outside world to be heard by others.

ICTs for climate change awareness

Recent progress in ICT and other technological innovations offers extraordinary opportunities for disaster preparedness, resilience-building and deepening connectivity. The highest number of fatalities from disasters per million people occur in the least developed countries. After the 2009 tsunami, the Samoan government implemented information dissemination via various media including AM radio, website and Facebook. As part of a tsunami early warning system, text messages were sent to village representatives who alerted villagers to evacuate by ringing school and church bells. Signage was introduced to ensure people knew where to go in the event of a disaster following a model introduced in New Zealand (UN-OHRLLS, 2017).

In Vanuatu drones are used for disaster assessment. The country is prone to various disasters from earthquakes, cyclones and volcanoes. Drones were used for disaster assessment in the aftermath of cyclone Pam and to determine possible evacuation routes in Ambae during volcanic activity. Small drones have also assisted in identifying areas that would be more vulnerable to tsunamis and to establish locations of the safest places. They are also useful for influencing development plans by determining the allocation of resources (UN-OHRLLS, 2017).

Policy makers recognise two key challenges associated with ICTs that can be categorised as infrastructure and content. Issues around infrastructure include high cost of towers and low return on investment because of dispersed population and rugged terrain; high cost of logistics; investment in ICT applications and services; and ensuring that investment in ICT benefits all. Challenges presented in the area of content include creating and managing relevant local content; lack of ideas concerning how to use ICT; limited online services; promoting more educational awareness across the country; and studying the pros and cons of regulating local content (UN-OHRLLS, 2017).

Media reporting

Mainstream media are seen as important channels for raising awareness. Key findings from the PACMAS Media and Communication Report 2013 highlighted how media reporting, and more specifically the language used by journalists, is a common issue and plays a key role in documenting climate change (Tacchi et al., 2013). Almost all respondents across the 14 Pacific countries in the study pointed out that the translation of scientific terms into local languages or in terms that communities can understand is challenging. Some of the findings from the PACMAS Report are shared below. While general awareness has been raised, journalists express difficulties in reporting interesting and new stories around the 'same' issues:

> I worry about the rural areas, where it's very basic, very rudimentary information about what climate change is, but they know – they recognize the changes – they just say it must be climate change but the full understanding of how one thing leads to the next – the cause and effect – that is what we try to do.

Fiji One TV encourages one of its journalists to specialise in climate change to help authorities convey their messages, bridging the jargon gap or technical term gap between scientist and journalist.

In some countries such as Kiribati and Palau, people report information fatigue with regard to climate change:

> It's hard to make climate change new and interesting. I mean we've been hearing about it for so long, that we need . . . some kind of a source that's providing us with new angle stories that we can kind of make into a local story.

This might be attributed to the difficulty of translating climate change issues into local stories that are relevant to people who might not see climate change as an immediate threat. A more effective way to communicate climate change would be to use local examples of environmental changes as suggested by an interviewee from Marshall Islands:

> You know, these are all things that people notice on their own. . . . The coastal erosion, the trees that are going down, the gravesites that are going down and that's where we point out and say 'that's climate change', that's an effect of climate change, and that's going to continue to happen. So those are the ways that we're able to connect with community members.

In Niue some participants expressed a need to acknowledge traditional ways to predict extreme weather patterns.

> Some people when they look out the sea, the birds, if the skies are a bit . . . orange . . . know there's a cyclone out there. We . . . start putting up our shutters even though the warning hasn't come out yet. Sometimes traditional knowledge actually helps us a lot more than some of the scientific stuff.

There is a propensity to report on events in the urban centre because it is cheaper and faster. Reporting from the field is important in order to share the work of scientists or community efforts in adaptation; however, resources are not always available for this to occur. With growing social media presence in the region, Facebook and, to some degree, Twitter posted by users, have become sources of information for journalists during natural disasters. For example, radio stations in Fiji and Vanuatu have turned to Facebook to receive updates from listeners in affected areas during floods and cyclones.

Regional communication networks

Regional organisations have realised the critical role of communication and are implementing a range of measures to enable information dissemination at various levels. These include developing an integrated approach to climate change policy and disaster risk management with due recognition of knowledge dissemination through education, training and awareness campaigns. Largely these use a one-way communication approach.

The Secretariat of the Pacific Regional Environment Programme (SPREP) has identified awareness as an important link in key climate change issues. SPREP has flagged several challenges at a regional level in relation to awareness programmes. These include attempts to ensure that scientifically correct and consistent messages and programmes are maintained; that they are site and community specific; and that they are less dependent on funding cycles (SPREP, 2012).

Various toolkits have also been developed by regional organisations. The Secretariat of the Pacific Community (SPC) has developed a gender and climate change toolkit for use by practitioners working on climate change programmes. The toolkit is designed as a guide to understand the links between gender and climate change in specific sectors such as food security, water, energy and disaster risk reduction by applying the 'gender lens' to relevant case studies, with a gender analysis checklist (SPC, 2013).

The Red Cross Red Crescent has developed a guidance note for its staff and volunteers on how to communicate climate change for risk reduction in the Pacific. Some of the communication principles it espouses include: using two-way communication; identifying trusted male and female champions; planning practical activities; starting with people's own lived experience; acknowledging and respecting local beliefs, values and practices.

Messages about climate change are disseminated at a regional level through various organisations and networks. *The Pacific Adventures of the Climate Crab* and *Cloud Nasara* are good examples of the edutainment model, in which short, animated films help increase awareness of the science and impacts of climate variability in the Pacific, with specific focus on how to prepare for future El Niño and La Niña events.

The Pacific Climate Change portal[1] acts as a hub for climate change information and knowledge sharing through resources such as an events calendar, links to documents, databases, an image and video gallery, and glossary definitions that are hyperlinked to key terms. Pacific Solutions Exchange,[2] a United Nations Development Programme (UNDP) listserv, encourages dialogue among its membership. Members post inquiries on specific topics and share answers, insights, experiences and lessons learned to help their counterparts in the Pacific. Fading Paradise is a multimedia resource developed by the Foundation of the Peoples of the South Pacific International to improve people's understanding about climate change. It is a compilation[3] of eight fiction stories by author Jim Bentley that portray the detrimental effects of climate change in the Pacific and are available in the form of comic books, audio CDs, 2D animation videos and Braille.

Community-led media

The communication challenges presented above point to the potential that ICTs offer in addressing issues of local participation and content that are inclusive of gender, youth and marginalised communities. As female-headed households are most affected by climate change impacts, Ospina and Heeks (2010) suggest that ICTs could play a key role by providing relevant information, capacity building, and empowerment to strengthen women's adaptive capacity.

Community radio is an important forum that invites participation by and gives voice to women and youth. femLINKpacific, a women-led community radio station in Fiji, has been documenting women's experiences with floods and other natural disasters in their communities since 2004. It uses rural networks such as women's clubs with 165 community-based women-led groups through their community media network. The Women's Weather Watch programme has a network of women weather reporters on the ground in different towns and villages who update listeners about the local condition and imminent threat to community from a rain event or other weather-related disasters. femLINKpacific campaigns for gender-based disaster risk management – response strategies "because women are often left out of the formal discussions and decision-making process in any disaster although they are the ones that are responsible at household level" (Tacchi et al., 2013 cited in Harris et al., 2016: 52). Its online presence includes Facebook, Twitter, Soundcloud and a website.[4]

Given the popularity of social media in the region, community-led media are more likely to make it easier for communities to actively shape and take part in their own awareness process, introduce their own solutions to climate change mitigation, and share these solutions with other Pacific communities. Project Survival Pacific, Fiji's youth climate change movement, is an example of youth-based social media, which mainly shares stories through its Facebook page[5] to reach out to young people, using approachable language. UNICEF Pacific's use of social media, in particular Facebook,[6] provides a better example of successful youth engagement in climate change communication. When its one-way communication-sharing platform resulted in poor youth engagement, they changed to a participatory approach that provided users with not only more opportunities to share and co-create, but also to control the Facebook page. This resulted in a 1,600% increase in engagement, including the youth-driven online game Pacific Climate Change, the objective of which is to explore social media as a platform for co-creating content.[7] This offers a good example of ICT use for interactive social dialogue, youth engagement and participatory communication.

Participatory media training projects have been implemented regionally as part of the PACMAS strategic activity on climate change awareness. They have adopted innovative participatory approaches that highlight local issues through community involvement. A notable project is the Action Against Climate Change (A2C2). It has produced a series of 21 videos as well as audio and online content to mentor high school students across the Pacific in the production of media content to raise climate change awareness and the understanding of how climate change will affect their lives and futures. This content has been disseminated through television, radio, newspapers and online platforms.[8]

The NGO 350.org, which coordinates global climate advocacy, offers its Pacific teams the opportunity to share their own perspectives on climate-related issues. Participatory media has become an important element of its work as it ensures stories about climate impacts coming out of the islands are shaped and told by Pacific islanders living with the impacts of climate change and not by someone else. This included their use of Twitter during tropical cyclones Pam in Vanuatu and Winston in Fiji. Small Voices Samoa is an initiative led by Brianna Fruean, a young ambassador who is combating climate change, using YouTube and Facebook to advocate for her school and community to take better care of their natural surroundings. She attributes her success to the use of participatory media through which she engages with her local and international community.

Two creative NGOs in Vanuatu have been using arts and theatre to educate communities. Wan Smolbag Theatre works at community level to raise awareness about a range of social and environmental issues (see case study 3.1 in Chapter 3). Through their travelling theatre groups and film productions, Wan Smolbag teaches about waste management, climate change and the ecological importance of turtles and reefs in Vanuatu and other Pacific Islands. Further Arts, a community arts organisation, runs a range of arts, media and cultural projects to develop long-term social and commercial enterprises that are culturally, socially, environmentally and financially sustainable. It uses participatory methodology in a peer-to-peer knowledge exchange through its media hub in Port Vila, Vanuatu.

Participatory approaches to storytelling are also being adopted by universities in the region. Students from University of the South Pacific (USP), are being trained to capture effects of climate change on their home regions using their mobile phones as part of an ePOP workshop. The ePOP (eParticipatory Observers Project) project was established by RFI Planète Radio, along with the National French Research Institute for Sustainable Development (IRD) to create a global network of youth who can capture changes in their environment by producing two- to three-minute videos. Bearing Witness climate change project, a Pacific Media Centre initiative, brings postgraduate journalism students from AUT University in Auckland and Pacific island communities together to produce stories of climate change. Journalism student Hele Ikimotu from Kiribati draws on his close connection to the issue and the Pacific region:

> Kiribati is one of the most affected countries by climate change. I have a special connection to the issues these communities

> are going through because it's my family that's being affected. I feel that mainstream media aren't doing enough to report on climate change. I think a project like 'bearing witness' gives a platform for climate change to be reported on genuinely and passionately, and gives opportunities to locals to tell their story. Rather than just talking about the need for change, I want to be a part of that change (Bell, 2018).

The participatory initiatives discussed in this chapter demonstrate that despite living on the margins of the digital revolution, Pacific island communities are proactively using media and communication to create awareness of and find solutions to the many environmental challenges.

Conclusion

Pacific island communities interact with various forms of ICT at the local level, using participatory approaches to develop content which is closer to their own realities and which reflects local cultures, values and individual aspirations. While ICTs offer possibilities for a younger generation, issues of illiteracy and capacity often remain overlooked. For these reasons more face-to-face and radio initiatives are favoured because they involve oral cultures and invite participation by those who cannot read or write.

Instead of single one-off projects that are sporadically funded, a more networked approach to participatory projects would encourage greater citizen engagement and improve community understanding of the issues. With assistance from governments and involvement of community-based environmental groups, a focused campaign that strategically incorporates participatory activities would incorporate local perspectives, cultural knowledge and concerns which are specific to the area.

It is important to gain greater understanding of the mediated implications of ICTs in climate change communication and disaster risk reduction at the community level. The short life span of participatory media activities means that the effectiveness of these projects is not properly evaluated. An evaluation tool, such as the DNA framework (see figure 2.1 in Chapter 2), will help to build a pool of knowledge for the implementation of future projects. This will bring greater understanding of the broader implications of using participatory environmental communication initiatives, which go beyond raising awareness. The framework provides opportunities for new alliances that engender cross-fertilisation of ideas and new ways of problem-solving. The process of local content production not only empowers the individual, it fosters greater engagement, reciprocity and social capital. These elements are important in the creation of more resilient communities.

Notes

1 www.pacificclimatechange.net
2 www.solutionexchange-un.net
3 www.pacmas.org/profile/fading-paradise-a-comprehensive-climate-change-resource-for-the-pacific/
4 www.femlinkpacific.org.fj/
5 www.facebook.com/projectsurvival
6 www.facebook.com/likeunicefpacific
7 www.facebook.com/likeunicefpacific#!/likeunicefpacific/app_130438327038952
8 www.pacmas.org/profile/action-against-climate-change-vanuatu/

References

ABC News. 2015. Nauru's president defends Facebook ban, says social media has "power to create instability". *ABC News*, 29 May 2015. Retrieved from www.abc.net.au/news/2015-05-29/nauru-president-baron-waqa-defends-facebook-ban/6507240

Bell, J. 2018. 'Bearing Witness climate storytellers gear up for fresh Fiji challenge', Asia Pacific Report. Retrieved from https://asiapacificreport.nz/2018/04/10/bearing-witness-climate-storytellers-gear-up-for-fresh-fiji-challenge/

Castells, M. 2000. *The Rise of the Network Society*. Oxford: Blackwell Publishers.

Cave, D. 2012. *Digital Islands: How the Pacific's ICT Revolution Is Transforming the Region* [Online]. Sydney: Lowy Institute for International Policy. Retrieved from www.lowyinstitute.org/files/cave_digital_islands_web.pdf

Chapman, R. & Slaymaker, T. 2002. *ICTs and Rural Development: Review of the Literature, Current Interventions and Opportunities for Action* [Online]. London, UK: Overseas Development Institute. Retrieved from www.odi.org/sites/odi.org.uk/files/odi-assets/publications-opinion-files/2670.pdf

Fuchs, C. 2009. A contribution to theoretical foundations of critical media and communication studies. *Javnost – The Public*, 16(2): 5–24. doi:10.1080/13183222.2009.11009001

Harris, U. S., Papoutsaki, E. & Kailahi, S. 2016. ICTs in climate change communication in the Pacific Islands. *Information Technologies & International Development*, 12(4): 47–58.

Heeks, R. & Ospina, A. V. 2018. Conceptualising the link between information systems and resilience: A developing country field study. *Information Systems Journal*, 1–27. doi:10.1111/isj.12177

ITU. 2017. *ICT Facts and Figures 2017* [Online]. Geneva, Switzerland: International Telecommunication Union. Retrieved from www.itu.int/en/ITU-D/Statistics/Pages/stat/default.aspx

Kalas, P. P. & Finlay, A. 2009. *Planting the Knowledge Seed-Adapting to Climate Change using ICTs: Concepts, Current Knowledge and Innovative Examples* [Online]. Melville, South Africa: Building Communication Opportunities (BCO) Alliance. Retrieved from www.apc.org/sites/default/files/BCO_ClimateChange.pdf

Khonje, Z. (2011). *Participatory Radio and the Public Sphere: The Case of Climate Change Communication in Malawi*. Retrieved July 18, 2014, from www.gen-derlinks.org.za/article/participatory-radio-and-the-public-sphere—the-case-of- climate-change-communication-in-malawi-2012-02-19.

Minges, M. & Stork, C. 2015. *Economic and Social Impact of ICTs in the Pacific* [Online]. Sydney, Australia: The Pacific Region Infrastructure Facility. Retrieved from www.theprif.org/index.php/resources/document-library/121-prif-ict-study-report-2015

Ospina, A. V. & Heeks, R. 2010. *Unveiling the Links between ICTs and Climate Change in Developing Countries: A Scoping Study* [Online]. Manchester, UK: Centre for Development Informatics, Insitute for Development Policy and Management, SED, University of Manchester. Retrieved from www.fao.org/fileadmin/user_upload/rome2007/docs/ICT%20and%20CC%20ScopingStudy.pdf

PRIF. 2015. *Economic and Social Impact of ICT in the Pacific* [Online]. Sydney, Australia: The Pacific Region Infrastructure Facility. Retrieved 21 January 2018 from https://theprif.org/index.php/news/53-media-releases/169-prif-ict-study

Sala, S. 2010. *The Role of Information and Communication Technologies for Community-based Adaptation to Climate Change* [Online]. Rome, Italy: Food and Agriculture Organization of the United Nations. Retrieved from www.fao.org/uploads/media/ap606e_2.pdf

SPC. 2013. Pacific gender & climate change toolkit. Retrieved on December 14, 2013, from: https://www.pacificclimatechange.net/document/pacific-gender-climate-change-toolkit-complete-toolkit

SPREP. 2012. Pacific climate change portal: Theme 4: Education, training and awareness. Retrieved on January 17, 2014, from: https://www.pacificclimatechange.net/projects

Tacchi, J., Horst, H., Papoutsaki, E., Thomas, V. & Eggins, E. 2013. *PACMAS State of Media and Communication Regional Report 2013* [Online]. Melbourne, Australia: PACMAS/ABC International Development. Retrieved from www.pacmas.org/profile/pacmas-state-of-media-andcommunication-report-2013/

The World Bank. 2017. *Long-Term Economic Opportunities and Challenges for Pacific Island Countries* [Online]. Washington, DC: The World Bank. Retrieved from http://documents.worldbank.org/curated/en/168951503668157320/pdf/ACS22308-PUBLIC-P154324-ADD-SERIES-PPFullReportFINALscreen.pdf

UN-OHRLLS. 2017. *Meeting Report: Regional Meeting for the Asia-Pacific Least Developed Countries* [Online]. Port Vila, Vanuatu: United Nations Office of the High Representative for the Least Developed Countries, Landlocked Developing Countries and Small Island Developing States. Retrieved from http://unohrlls.org/custom-content/uploads/2017/11/Regional-Meeting-for-Asia-Pacific-LDCs-and-Pacific-SIDS-Broadband-Connectivity_LowRes1.pdf

Yap, N. T. 2011. *Disaster Management, Developing Country Communities & Climate Change: The Role of ICTs* [Online]. Manchester, UK: Centre for Development Informatics, Insitute for Development Policy and Management, SED, University of Manchester. Retrieved from www.niccd.org/wp-content/uploads/2017/11/YapDisasterManagementDevelopmentICTs.pdf

7 Who is listening?

Reflection

As I write this chapter, a Category 5 Tropical Cyclone, Gita, is sweeping through the South Pacific islands of Samoa and Tonga, leaving a trail of destruction. Using the hashtag #TCGita I learn that the Tongan Meteorological office building was damaged and can no longer send updates about the approaching cyclone. That function has transferred to the Fiji Meteorological Service. The Parliament House is also completely destroyed. Next morning, I turn on Australian television for updates on Tonga's fate from TC Gita. Instead, the top story is about a politician's fate arising from his infidelity. I return to Twitter to find a post from Tonga: "It's 4.20am now and what a freaking night. They can never portray what this tropical cyclone is in any movie". Responses from family and friends to the tweet express relief that the family is safe. Someone is listening.

By taking over the reins of production, people gain control of their own storytelling and by implication their own lives. Climate change is a crosscutting issue that is making its presence felt at the intersections of culture, politics, economy and ecology. With natural disasters increasing in regularity and intensity, a range of information and communication tools is necessary as a lifeline to emergency assistance and long-term sustainability of vulnerable communities. Participatory environmental communication (PEC) opens up shared spaces for modern-day storytelling, enabling a shift from binary frameworks towards a more holistic and networked communication model.

From international agencies to local NGOs, from scholars to practitioners, there is a growing recognition for need to change the way in which we see the world and the concepts we use to understand it. The importance of transdisciplinarity, and a need to invite wider perspectives are part of this process. One of the ways this transformation can be facilitated is through the triad of theory, experiential learning, and reflexive research.

Participatory environmental communication, undergirded by the DNA (*diversity, network, agency*) framework, offers such a systems approach to problem-solving. The approach integrates interdisciplinary knowledge, inspires collaboration and dialogue at all levels, and utilises information networks to catalyse the agency of ordinary people towards collective action. It also includes systems of knowledge that have been silenced. These are the use of folk songs and stories, or the powerful acts of protest that activists and indigenous groups deploy in environmental movements discussed in earlier chapters.

This final chapter urges scholars, practitioners and activists to include participatory environmental communication strategies in their toolbox of community engagement practices on environmental issues. It argues for increased adoption of various forms of participatory media and communication as a way of ensuring greater diversity of voices at the local level. To function effectively, participatory strategies require the cooperation of diverse actors by enlisting vertical and horizontal communication networks. This includes politicians and policy makers who have the power to make policy decisions. It needs the expertise of scientists, special service providers, first responders and field workers who are able to provide the necessary support to build the capacity of vulnerable communities. Above all, participatory strategies must enable citizens to share their experiences and influence policy decisions so that frontline communities are empowered to have control over their own lives.

Linking the threads of cooperation

Citizen voices have struggled to summon sustainable communicative technologies and media forms that enable uncorrupted representation of their views. As every click of our computer captures and defines our identity in algorithms, we are forced towards a further narrowing of our choices. If human evolution has been dependent upon collective learning using networks which share information and knowledge (Christian, 2011), then centralisation of communication structures that benefit some and keep out others would seem counter-productive.

Individual change is proving to be insufficient by itself. To enact change at a global level "community dialogue, collective, cooperative action is required" (Figueroa et al., 2002: 3). Communication builds trust and "plays a critical role in the process of collaboration" (Rogers et al., 2016: 79) between cross-sectoral stakeholders such as government and non-government organisations, knowledge holders and community groups.

The spread of information is analogous to the flow of water. Imagine for a moment how water finds its natural course through unseen crevices, sloping pathways, streams and rivers, while at times being blocked and diverted or quietly disappearing into the parched earth. Participatory mapping of information flows in a society identifies how information is received, shared and used, including impediments to the flow. These information flows help to identify trusted sources in the local context vital during emergencies (see Table 3.2 in Chapter 3). It is a critical first step in designing participatory environmental communication projects.

Communities are unlikely to respond to government policies promoting mitigation and adaptation strategies without an improved perception of climate change risk at a local level. Mapping the flow of information between decision-makers and citizens is an important part of this process. What are the relevant channels of communication? Who disseminates the information? What impact is the information having on the general population? These are questions for both practitioners and researchers working in the area of climate change.

The main aim of the DNA framework is to reveal connections between people and organisations and to gauge how well these connections enable information flows, i.e. how, where and with whom people exchange information, and the quality of these exchanges within a given context. It is a critical first step in designing PEC projects. This prepares both individuals and communities during times of crisis and builds resilience both at individual and community levels. This participatory approach shifts the climate change discourse away from predominantly one-way expert-led elite perspectives, towards a dialogic model, which gives voice to frontline communities. Elizabeth Holland, director of the Pacific Centre for Environment and Sustainable Development (PaCE-SD) at the University of the South Pacific echoes this sentiment: "Local research in the region is of utmost importance, leading to the informed development choices and is the best way forward as it creates a direct connection between the research and the communities once it is implemented" (Robie, 2018).

Participatory environmental communication

Throughout the process of writing and researching this book, I have become acutely aware of an emerging consciousness that seeks another, more inclusive and holistic approach that values dialogue, collaboration and interconnectedness among diverse entities. Nisbet and Scheufele (2009: 1767) state: "One can detect a growing recognition that effective communication requires initiatives that sponsor dialogue, trust, relationships, and public participation across a diversity of social settings and media platforms". Social and economic theorist, Jeremy Rifkin, in his book *The Third Industrial Revolution* (2011), argues that for a system to be resilient it needs to be distributed.

The DNA framework introduced in chapter 2 offers a theoretical tool that fills an important gap in communication between disparate and divergent groups. *Diversity*, in the form of collaboration among various entities, enables innovative and transformative thinking; *Networks* form complex webs of relationships vital for information exchange and survival of an ecosystem; *Agency* is the capacity of all beings to act upon their environment in order to bring about change.

The DNA framework is highly contextual. As such, it can be used for a variety of purposes to identify opportunities for collaborative engagement across individuals, organisations and programmes, as identified below:

1. To link heterogeneous networks bringing diverse actors into a collective
2. To monitor and evaluate PEC projects to ensure that *DNA* elements are present in resilience-building programmes
3. To analyse participatory environmental communication projects using the *DNA* essentials (see table 2.1 in chapter 2).

Diversity in PEC

While it is known that diversity encourages innovation and out-of-the-box problem-solving, it is still not fully accommodated in resilience-building activities for climate change adaptation and disaster risk reduction. Ilan Kelman (2017) argues that climate change, disaster risk reduction and sustainable development planning need to be better linked. He found parallels in all three areas, yet no effort is made to interlink these programmes because of an entrenched organisational culture that has developed around each. For example, "Flood, weather and drought risk reduction enacted for climate change adaptation is exactly the same as being enacted for wider disaster risk reduction" (2017: 257). He suggests an integrated approach which connects all three areas should be adopted for planning future activities. To bridge the divide and facilitate collaborative engagement across organisations, diversity must be built into the programme cycle at the beginning. The diversity matrix is a useful tool for mapping diversity (see table 3.1 in chapter 3) which can be revised to fit into the needs of an individual, community or organisational setting.

Participatory environmental communication invites diversity of voices, values and beliefs, as well as cultural perspectives. It uses a wide variety of media forms and communication channels such as face-to-face communication, folk media, traditional and new media. Digital media encourages intergenerational knowledge exchange between young people and elders of a community. Many young people are leaving traditional ways of life to live and work in urban areas and losing interest in old ways. This distance means that old ways of knowledge exchange where elders in the community progressively taught the young about traditional knowledge are no longer possible. With the passing away of the elders, traditional ecological knowledge, such as farming, fishing and food preparation, is being lost. By introducing digital media in adaptation programmes, young people are more enthusiastic about working with elders as seen in the case study of Andra Islanders in chapter 4. In many traditional societies, their knowledge cannot be shared with outsiders, so involving younger members of the same community ensures that the project observes ethical standards. This two-way exchange of intergenerational knowledge and skills gives both the young and the old in the community equal partnership, as young people use their knowledge of digital media to record traditional knowledge of their elders.

Network in PEC

Local area networks (LAN) is a computing term which describes a network of interconnected devices that exist in one location such as university, hospital or one building, so that users can share resources. This term can also be used to describe the participatory communication networks in a local area.

When mainstream flows of information are disabled during emergencies, people proactively search out local networks to share information and resources in their vicinity. People's patterns of communication revert to trusted word of mouth networks or ICTs that provide immediate access to crucial services. With electricity and communication infrastructures damaged, older forms of communication prove more reliable. Internews' analysis of the changing information landscape in New York City during Hurricane Sandy found ham (amateur) radio and corded landlines replaced mainstream media and cellphones, due to power outages and downed mobile towers (Internews, 2014). During emergencies safe public spaces, places of worship and community halls become the central node of information exchange.

Interpersonal and local communication networks have proved more relevant than mainstream media during times of crisis because these networks seek out and provide information that fulfill the demand in the local context. An analysis of disaster information needs during the reconstruction phase following the 2004 Tsunami in Aceh, Indonesia (Yusuf, 2017), and again during Hurricane Sandy in New York (Internews, 2014) demonstrate that mainstream media was less reliable as a source of real-time information because they used conflict-event framing devices, thereby focusing on sensational reports instead of servicing community needs. As such, mainstream media should be considered as only one aspect of the rich information ecosystem that already exists within communities, such as social media, community media, and face-to-face exchanges. Of these, face-to-face communication is the most powerful during crisis. Mapping community information flows makes visible the patterns of communication that only local people are aware of within a community. Table 3.2 (Chapter 3) identifies community-based information flows for a local area. It does not include any outside forms of media, such as mainstream radio and television. This local area network map, which isolates only local information exchanges, creates a resource for the community in times of disaster when all outside communication is lost.

Agency in PEC

Broadening public engagement is critical in solving environmental challenges. This research has established that effective communication occurs when people share a collective vision for change. Communication is a two-way process, which requires active listening of voices in the margins, in order to enact real change at ground level. Dialogue creates networks and brings diverse actors to the table.

Dialogue is more than just talk. Participants uncover the layers of complexity within their own lives, their community and the wider world when they are in dialogue with others. It helps them to recognise what is happening now, and why. This powerful moment of realisation invariably begins with a face-to-face encounter. At this critical juncture, conversation can turn into a debate that divides, or a dialogue that engages participants towards a common purpose. A facilitator's role is important here to assist people to find their voice and a common purpose.

Dialogue thus is the bedrock of communication and brings people together towards a common goal. Through dialogue and talanoa sessions (Fijian dialogic method) people begin to identify the links between environmental changes and impacts on their livelihoods, access to energy, food and water security, health and sanitation. Karin Gwinn Wilkins (2011: 389) notes: "A dialogic model values media for their potential to inspire individual reflection and collective action. Dialogue might help people with similar concerns build cohesive strategies toward resolving problematic conditions". When people recognise their own commitment to an issue, it is then that they are truly inspired to action.

These portrayals are absent in mainstream media, where the hegemonic debates of dysfunctional power elites dominate the daily news headlines. Rodríguez's theory of citizen's media as a dynamic site for action and transformation of ordinary citizens also aligns with notions of agency (Rodríguez, 2001). Rodríguez recognises this agency of community media producers, when she argues for the reframing of the theoretical discourse from alternative to 'citizens' media': "However, despite their geographic, economic and cultural differences, they all have one thing in common: they express the will and agency of a human community confronting historical marginalizing and isolating forces, whatever these may be" (2001: 63).

Throughout this book the importance of communication has been a consistent theme. From scientists to policy makers, from communities to first responders, all recognise that without good communication, progress is seriously hampered in both risk communication and emergency service delivery. To have good communication individuals, communities and organisations must have a clear understanding of how information flows within their local area and across sectors. A clear plan of action demands collaboration in establishing communication protocols. This means dialogue, not dissemination. Dialogue takes into account the other parties' needs and highlights the context in which information is used. During emergency situations a clear and consistent message is critical. How the messages are understood and used is dependent on the local context in which it is received, including the ability of people to trust the message and the message givers.

Public participation in protest marches, environmental movements and advocacy campaigns has contributed significantly to enacting policy changes and raising awareness about environmental issues in contemporary society. A new way of inviting participation is now necessary in order to drive change at a personal and community level. Such a response requires engaging communities through prevailing community networks and accessing diverse forms of knowledge. This would include scientifically reliable information delivered within the context of local knowledge that makes sense to people in their daily lives. Participatory Environmental Communication that weaves three widely used concepts, *diversity, network and agency*, offers a cohesive underlying system to bring scientists, practitioners and diverse communities together in a dialogue about pressing environmental matters.

A final reflection

> Our lonely blue planet is like an island in space. Like an island, the Earth is both remote and interconnected, feeding off the energy of the sun while its gravitational pull is in constant dance with the moon. Yet it is a delicate system with finite resources. Let us find a common purpose through dialogue and collaboration to ensure the wellbeing of our common home for both present and future living beings.

References

Christian, D. 2011. *Maps of Time: An Introduction to Big History*. Berkeley, CA: University of California Press.

Figueroa, M. E., Kincaid, D. L., Rani, M. & Lewis, G. 2002. *Communication for Social Change: An Integrated Model for Measuring the Process and Its Outcomes*. New York: The Rockefeller Foundation and Johns Hopkins University Center for Communication Programs.

Internews. 2014. *Information Ecosystems in Action: New York* [Online]. Arcata, CA: Internews. Retrieved from www.internews.org/sites/default/files/resources/Internews_Information_Ecosystems_in_Action-NewYork.pdf

Kelman, I. 2017. Linking disaster risk reduction, climate change, and the sustainable development goals. *Disaster Prevention and Management: An International Journal*, 26(3): 254–258. doi:10.1108/DPM-02-2017-0043

Nisbet, M. C. & Scheufele, D. A. 2009. What 's next for science communication? Promising directions and lingering distractions. *American Journal of Botany*, 96(10): 1767–1778. doi:10.3732/ajb.0900041

Rifkin, J. 2011. *The Third Industrial Revolution: How Lateral Power is Transforming Energy, the Economy, and the World*. New York: Palgrave Macmillan.

Robie, D. 2018. More frontline research "by Pacific for Pacific" plea at climate summit. *AsiaPacificReport*, 25 February 2018. Retrieved from https://asiapacificreport.nz/2018/02/25/more-frontline-research-by-pacific-for-pacific-plea-at-climate-summit/

Rodríguez, C. (2001). Fissures in the Mediascape: An International Study of Citizens' Media. Cresskill, NJ: Hampton Press.

Rogers, P., Burnside-Lawry, J., Dragisic, J. & Mills, C. 2016. Collaboration and communication: Building a research agenda and way of working towards community disaster resilience. *Disaster Prevention and Management*, 25(1): 75–90. doi:10.1108/DPM-01-2015-0013

Wilkins, K. G. 2011. Participatory media. *In:* Downing, J. D. H. (ed.) *Encyclopedia of Social Movement Media*. Thousand Oaks, CA: Sage Publications.

Yusuf, I. A. 2017. *Aceh Community Radios in Disaster Information Management: Risk Reduction, Resilience and Preparedness*. The 3rd World Conference on Media and Mass Communication 2017 (MEDCOM 2017), 20–22 April, 2017, Kuala Lumpur, Malaysia.

Index

Note: Page numbers in *italics* indicate figures and in **bold** indicate tables on the corresponding pages.

For Product Safety Concerns and Information please contact our EU representative GPSR@taylorandfrancis.com Taylor & Francis Verlag GmbH, Kaufingerstraße 24, 80331 München, Germany